MICROWAVE TRANSITION DESIGN

MICROWAVE TRANSITION DESIGN

Jamal S. Izadian
Shahin M. Izadian

Artech House

Library of Congress Cataloging-in-Publication Data

Izadian, Jamal S.
    Microwave transitions and applications.
    Includes bibliographies and index.
    1. Micirowave transmission lines. 2. Impedance matching. I. Izadian,
Shahin M. II. Title.
TK7876.I92   1988   621.381'3   88-6342
ISBN   0-89006-235-8

International Standard Book Number: 0-89006-235
Library of Congress Catalog Card Number: 88-6342

10   9   8   7   6   5   4   3   2   1

DEDICATION:

To our parents,
our sons Shervin and Shyan,
and all other children to whom the future belongs.

Contents

Preface

We have tried to bring together all aspects of microwave transition design to help the microwave circuit and subsystem designer to save time and to finish their design by putting it into another medium (i.e., the microstrip circuit in waveguides, or devices and MMICs in packages). We designed this book as the only reference that answers the question, "How do we make a microwave transition." We have included enough introductory material, along with information on practical transitions, that the designer need not conduct widely scattered research in order to understand how a particular transition works and how to design it. We hope that this book will teach what has been so obscure by putting the subject in the perspective of the electromagnetic fields, tied to the physical impedance concept. We believe that the presentation should lend itself to improving microwave design and measurement by including all the transition and fixture effects, using CAD models and de-embedding or embedding.

In Chapter 1, we present an introduction to the microwave transition concept by providing a general definition to describe a microwave transition. The impedance concept and its various definitions are presented to clarify some ambiguities that exist in practice. A discussion of tapered transitions is presented with an analogy to nonuniform transmission lines, in which the theory of the Fourier transform can be used to synthesize a given reflection coefficient by synthesizing an appropriate impedance taper profile.

An important concept is presented and emphasized throughout the book: in making a transition, a field match is the prerequisite to a proper impedance match, and the fields of the two media must be made similar. Therefore, the mechanical transition provides the field match, whereas the electrical design provides the impedance match. Finally, in this first chapter, a coherent procedure is given to aid the design of any microwave transition. Such a unified approach has not been previously suggested, and a transition has until now not been so coherently presented.

ix

Chapters 2 to 5 present various microwave transitions. In Chapter 2, several approaches to design and characterization of coaxial-to-microstrip lines are given. In this chapter, models for the transition are provided and techniques to extract the model element values from a series of measurements are suggested. This chapter will prove to be an invaluable tool for better transition and circuit designs. This is because we provide techniques for the modeling of a transition so that it can be included as a circuit file in any microwave CAD program. This makes it possible to model the design to account for the transition effects.

Chapter 3 presents the waveguide-to-coaxial line transition. The most common approaches are illustrated, including classical approaches and some newer techniques. In this chapter, there is a detailed discussion and formulation of the various techniques so that the chapter is self-contained for the development of CAD tools. A detailed representation of the work of various researchers has been compiled to make the chapter a complete reference, including a circuit model for the transition that will be useful for microwave CAD programs.

In Chapter 4, we provide the waveguide-to-microstrip and fin-line transitions in detail. Several useful approaches are discussed, including ridged waveguide and antipodal fin line. In each case, the field match evolution is illustrated in the figures to describe the mechanical approach of the field match and the electrical design of the impedance synthesis.

Chapter 5 presents additional transitions, which are very useful, but do not come under the categories covered in previous chapters. These transitions are mostly between the planar transmission lines themselves, such as slotline transitions, coplanar waveguide-to-microstrip or slotlines, a new transmission line called microslab, and the dielectric waveguide.

Chapters 6 and 7 give microwave test fixtures, and their de-embedding and characterization, respectively. The test fixture is presented in detail in Chaper 6, mainly because this continues to be both the most important and perhaps the most neglected part of a very sensitive measurement system. The test fixture is seldom characterized correctly because it is usually difficult and more involved than most working engineers care to admit. In this chapter, a unified and coherent procedure for designing a proper test fixture is first presented. Then some actual designs are suggested for coaxial, waveguide, and fin-line test fixtures. The need for providing biasing network and control lines is also emphasized. Suggestions for the design of a proper chip carrier for testing transistors or MMICs are also given as well as a method of calibration for the fixtures.

De-embedding is considered to be an essential part of microwave transition modeling. A transition must be properly characterized to enable the designer to account for its effects on overall circuit performance. This is especially useful for test fixture calibration and de-embedding or embedding the transition effect in the measurement data. Some popular techniques, such as TRL, LRL, TSD, are presented in this chapter. A discussion of possible application of time-domain de-embedding is also given. An application in device modeling is provided to show the usefulness and the significance of de-embedding in addition to a discussion of its extension to wafer probing.

Finally, the book would not have been complete without illustrating the transition to antennas. Chapter 8 presents several examples and insights into antenna design from the viewpoint of the transition. We show that the problem is not really different than that of the circuit design, and the same techniques apply equally well.

The authors are pleased to thank Dr. Alfred N. Riddle, who painstakingly read the first draft of the manuscript and gave very valuable suggestions. Also thanks are due Mr. David Cahana, who very patiently and meticulously read the second draft and gave excellent and insightful suggestions. Our thanks also go to our two sons, Shervin Daniel and Shyan David, for their sacrifices. They sometimes went to bed early to let us work on this project. Finally, we want to thank a person who has always inspired us to expect and to achieve excellence, the co-author's brother, Dr. Jalal Izadian. We salute you, professor.

Chapter 1
Introduction

1.1 INTRODUCTION

A microwave transition is the mechanism by which the electromagnetic wave on one type of a transmission line is coupled into another type, such as a transition between waveguide and microstrip line, or waveguide and coaxial line.

A transition between two different media is one that transforms the electromagnetic field configuration to conform with the boundary conditions presented by the new transmission line. The objective is to make this transformation as efficient as possible.

For a better understanding of the transition mechanism between two media, it is helpful to distinguish between impedance and the field transitions. From the viewpoint of impedance, the transition between two transmission media must provide an impedance matching between the two transmission lines that maximizes coupling while minimizing reflections. It is essential to use an impedance definition that is suitable for the transmission line under consideration. As will be shown later, several impedance definitions used in practice are not consistent, and further clarification needs to be made.

In addition to the impedance match, and as a prerequisite, a transition must provide an efficient field transition from one medium to another by smooth and gradual change in the physical boundary conditions. This gradual field match is a function of the change in the boundary conditions as the transmission line changes shape (e.g., from a waveguide to a ridged waveguide to a microstrip, *etcetera*), whereby the field geometry will change accordingly [1]. To achieve both a field match and an impedance match, step transition or continuous taper transition is often used. Before considering any transition design, it is necessary to clarify further what is meant by field match and why there are some ambiguities in the definition of impedance.

1

1.2 FIELD MATCHING BETWEEN TRANSMISSION LINES

It is well known that the time-harmonic electric and magnetic fields are coupled as a result of Maxwell's equations. Furthermore, Maxwell's equations state that the electric and magnetic field lines are always orthogonal. Therefore, it suffices for our discussion to consider only one field, either electric or magnetic, because the information for the other field is readily known.

The electromagnetic boundary conditions on a *perfect electric conductor* (PEC) state that the tangential electric field on the surface of a PEC vanishes. This simply implies that the electric field is perpendicular to the surface of the PEC. If the conductor is not a perfect conductor, if it has a finite conductivity, then there is a slight tangential electric field, which implies that the electric field is not quiet perpendicular to the surface.

By duality, a similar boundary condition applies to the magnetic field on the surface of a *perfect magnetic conductor* (PMC). It should be kept in mind that the PMC is a tool of convenience, because in reality there is no such thing. On a PEC, the tangential magnetic field converts to the surface currents. The mathematical expressions stating the above boundary conditions are

$$\hat{n} \times (E_2 - E_1) = 0 \tag{1.1}$$

$$\hat{n} \times (H_2 - H_1) = J_s \tag{1.2}$$

where J_s is the electric surface current density, and $\hat{n}$ is the surface normal pointed out of medium 2 away from medium 1.

Equation (1.1) states that the tangential component of the electric field vanishes at the surface of a PEC or that the field is normal to the surface of PEC, and equation (1.2) states that the tangential component of the magnetic field on a PEC converts to surface currents with its direction and density determined by the direction and magnitude of the magnetic field. Figure 1.1 shows some typical examples of the boundary conditions illustrating the electric and magnetic field geometries, such as the coaxial line, the microstrip, the rectangular waveguide, ridged waveguide, unilateral and antipodal fin lines, and convex surfaces.

The field geometry match is thus achieved by providing a reshaped physical structure to provide a new boundary condition reshaping the field

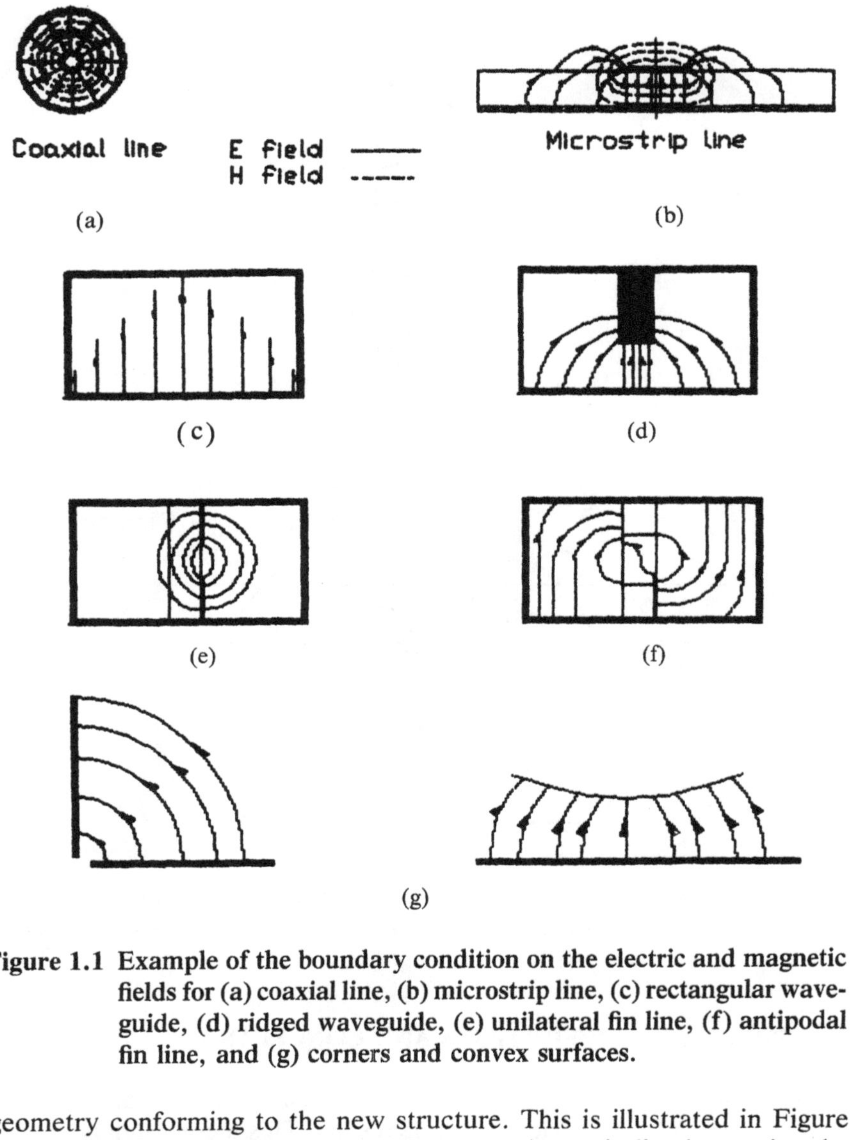

Coaxial line   E field —————
H field ------

Microstrip line

(a)

(b)

(c)

(d)

(e)

(f)

(g)

Figure 1.1 Example of the boundary condition on the electric and magnetic fields for (a) coaxial line, (b) microstrip line, (c) rectangular waveguide, (d) ridged waveguide, (e) unilateral fin line, (f) antipodal fin line, and (g) corners and convex surfaces.

geometry conforming to the new structure. This is illustrated in Figure 1.2, in which a coaxial line is changed to a microstrip line by cutting the coaxial line along the longitudinal direction and unfolding it to evolve into the microstrip line having a new boundary condition and new field geometry.

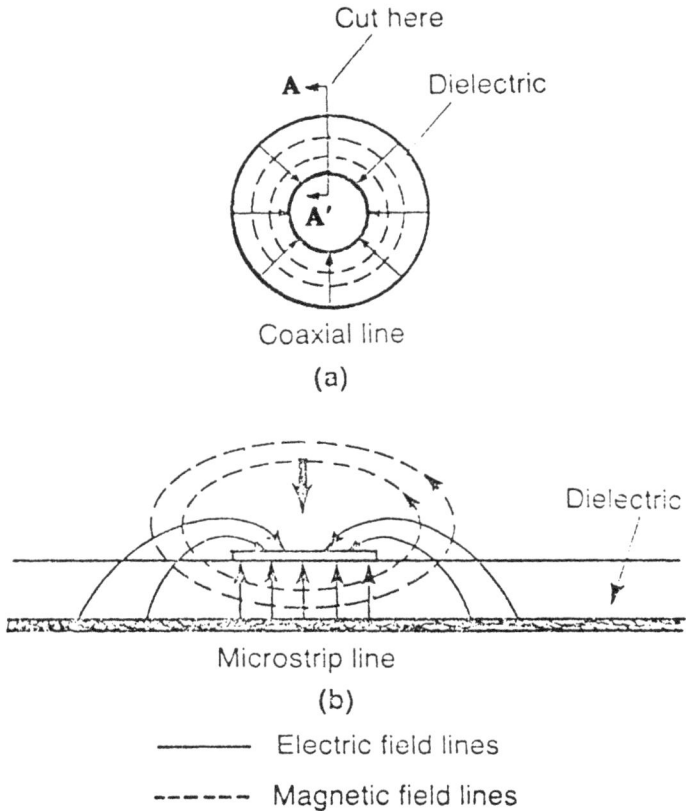

Figure 1.2 Illustration of evolution of a microstrip line from a coaxial line by cutting the coaxial line along a longitude (A–A') and unfolding.

1.3 IMPEDANCE DEFINITIONS AND CONCEPTS

There are at least three different ways of defining impedance, and generally they will not provide the same results. This inconsistency between the result and the impedance definition has been a great source of controversy among practitioners, mainly because anyone can justify the use of his or her favorite definition and find adequate arguments against the others. No significant scientific attempt has been made as yet to come up with a standard. An attempt to explain this inconsistency has been made by Brews [2], and a partial result will be presented here to help clarify some of the ambiguities in the impedance definitions.

Considering a waveguide as an equivalent transmission line, the equivalent voltages and currents are given by

$$E(x,y,z) = V_e(z)E(x,y) \tag{1.3a}$$

$$H(x,y,z) = I_e(z)H(x,y) \tag{1.3b}$$

where $V_e(z)$ and $I_e(z)$ are the voltages and currents on the equivalent transmission line, and the fields are transverse to Z.

The ratio of $V_e(z)$ to $I_e(z)$ is called the *modal wave impedance:* $Z_e = V_e/I_e$, where

$$Z_e = \sqrt{\frac{\mu}{\epsilon}} \frac{\lambda_g}{\lambda} \text{ (TE mode)} \tag{1.4a}$$

and

$$Z_e = \sqrt{\frac{\mu}{\epsilon}} \frac{\lambda}{\lambda_g} \text{ (TM mode)} \tag{1.4b}$$

where λ_g is the mode-dependent guide wavelength.

Additionally, the waveguide current and voltage, I_g and V_g, usually are defined by

$$V_g^{(z)} = \int_a^b E \cdot dL \tag{1.5a}$$

$$I_g^{(z)} = \int_c H \cdot dL \tag{1.5b}$$

where a and b are two points on the cross section where the voltage difference is needed, and the c is the closed contour enclosing a conductor on which the current is sought. Contour c is usually in the plane containing a and b on cross section.

Three definitions for the *waveguide impedance* are voltage-current, power-current, and power-voltage relationships, which are given by

$$Z_{VI} = \frac{V_g}{I_g} \tag{1.6a}$$

$$Z_{PI} = \frac{2P}{|I_g|^2} \tag{1.6b}$$

$$Z_{PV} = \frac{|V_g|^2}{2P} \tag{1.6c}$$

where

$$P = V_g I_g^*/2 \qquad (1.7)$$

and is not equal to the longitudinal component of the $S = (E \times H^*)/2$ integrated over the cross section of the waveguide. This is one reason why the definitions will not agree, as will be discussed later. (Note that the asterisk here represents complex conjugation.)

For the dominant TE_{10} mode the three definitions of the waveguide impedances become [3]

$$Z_{VI} = Z_e \frac{\pi b}{2a} \qquad \Omega \qquad (1.8a)$$

$$Z_{PV} = Z_e \frac{2b}{a} \qquad \Omega \qquad (1.8b)$$

$$Z_{PI} = Z_e \frac{\pi^2 b}{8a} \qquad \Omega \qquad (1.8c)$$

$$I_g^{(z)} = I_e^{(z)} \frac{2}{\pi} \sqrt{\frac{2a}{b}} \qquad (1.8d)$$

and

$$V_g^{(z)} = V_e^{(z)} \sqrt{\frac{2b}{a}} \qquad (1.8e)$$

where I_g and V_g are the waveguide currents and voltages and V_e and I_e are the equivalent transmission line currents and voltages, respectively. Brews [2] equates the power defined by circuit theory to that defined by the Poynting theorem to clarify impedance.

$$P = \frac{V_g I_g^*}{2}$$

$$= \frac{1}{2} \int \int (E \times H^*) \cdot ds = \frac{1}{2} \int \int (E_x H_y^* - E_y H_x^*) \, dx \, dy \qquad (1.9a)$$

This equality holds if and only if

$$\iint (E_{tx}H_{ty}^* - E_{ty}H_{tx}^*)\,dx\,dy = 1 \tag{1.9b}$$

$$\iint \left(\frac{1}{\mu}|(\nabla \times E_t)|^2 - \omega^2\epsilon|E_t|^2\right)dx\,dy = j\omega\gamma/Z_0 \tag{1.9c}$$

and

$$\iint \left(\frac{1}{\epsilon}|(\nabla \times H_t)|^2 - \omega^2\mu|H_t|^2\right)dx\,dy = j\omega\gamma Z_0 \tag{1.9d}$$

are satisfied. This can easily be deduced by using the following expressions for the electric and magnetic fields separated into transverse and longitudinal components and equivalent transmission line currents and voltages:

$$E(x,y,z) = E_t(x,y)V_e(z) + \eta E_1(x,y)I_e(z) \tag{1.10a}$$

$$H(x,y,z) = H_t(x,y)I_e(z) + H_1(x,y)V_e(z)/\eta \tag{1.10b}$$

where the voltage and currents satisfy the transmission line equations:

$$\frac{dI_e(z)}{dz} = -\frac{\gamma}{Z_0}V_e(z) \tag{1.11a}$$

$$\frac{dV_e(z)}{dz} = -\gamma Z_0 I_e(z) \tag{1.11b}$$

where

$$V_e(z) = Ae^{-\gamma z} + Be^{\gamma z} \tag{1.12a}$$

$$I_e(z) = (Ae^{-\gamma z} - Be^{\gamma z})/Z_0 \tag{1.12b}$$

These equivalent voltage and current are now given by

$$I_e(z) = \left(\frac{1}{j\omega\gamma Z_0}\right)^* \iint \left\{H(x,y,z) \cdot \left\{\nabla \times \left[\frac{1}{\epsilon}\nabla \times H_t\right] - \omega^2\mu H_t\right\}^*\right\}dx\,dy \tag{1.13a}$$

$$V_e(z) = \left(\frac{Z_0}{j\omega\gamma}\right)^* \int\int \left\{E(x,y,z) \cdot \left\{\nabla\right.\right.$$

$$\left.\left.\times \left[\frac{1}{\mu}\nabla \times E_t\right] - \omega^2\epsilon E_t\right\}^*\right\} \, dx \, dy \qquad (1.13b)$$

A comparison of these expressions with the ones given previously in (1.3) and (1.5) using arbitrary contour integrations reveals that, in order for all three definitions of impedance to agree, the voltage and currents must be obtained using the more general expression (1.13). Furthermore, it is obvious that the new generalized expressions for equivalent voltage and current are coupled due to the condition given by (1.9). This means that defining either voltage or current is sufficient for complete specification of an impedance that will be consistent with all the three definitions.

In concluding this section, it must be said that, for a single mode propagation waveguide (general), condition (1.9a) must be met to ensure that all the impedance definitions are equivalent. Equation (1.13) thus provided more general expressions for the equivalent voltages and currents that are coupled due to equations (1.9c) and (1.9d).

A word of caution is still in order. The above discussion basically related all definitions due to consistency of power definitions, but the choice of which impedance definition is to be used must still be decided by the practicality of the problem at hand. The author suggests that consistency in definition of impedance for all media in the transition is still the key to success, with, of course, use of basic common sense.

1.4 IMPEDANCE TAPERS

Tapered transmission lines are used to match two transmission lines of different characteristic impedances. The transition can be made using cascades of many quarter-wave transformers or a continuous taper. The step transition has been presented in depth by Cohn [4], where he provides procedure for realization of an optimum Chebyschev stepped transition between two lines of unequal impedances.

The Chebyschev transformer provides an optimum equal ripple reflection coefficient response. A maximally flat response impedance transformer can be realized by a binomial expansion technique, as outlined in Collin [5]. The details of these techniques will not be presented here for the sake of brevity; however, the interested reader can consult the literature for more details.

A continuous taper impedance transformer is shown in Figure 1.3 which is essentially a piece of a nonuniform transmission line. The reflection coefficient of a nonuniform transmission line, ρ, is a function of

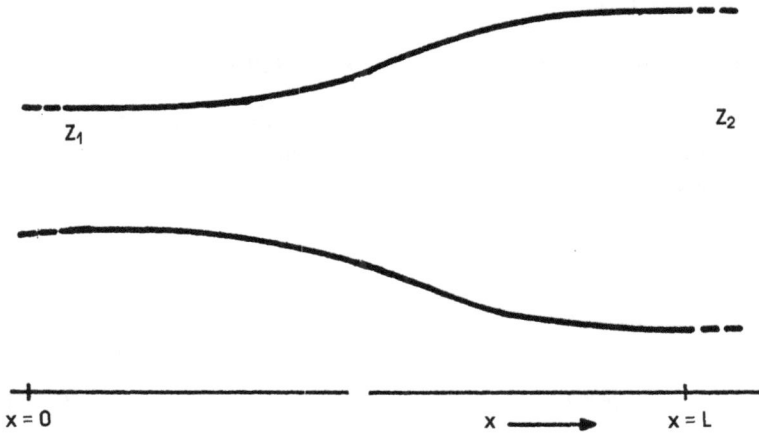

Figure 1.3 A continuous impedance taper is essentially a nonuniform transmission line of length L.

position along that line, which can be found from a nonlinear differential equation known as the Riccati differential equation, given by Bolinder [6]:

$$\frac{d\rho}{du_1} = \frac{1}{2}\frac{d(\ln Z_0)}{du_1}(l - \rho^2) - 2\rho = 0 \tag{1.14a}$$

where Z_0 is the characteristic impedance along the line, and

$$u_1 = \int_0^x \gamma\, dx \tag{1.14b}$$

where γ is the propagation constant along the line $= \alpha + j\beta$. There are some alternative forms of this equation in terms of the impedance [7]. Solutions to this differential equation are not readily available without some assumptions to simplify the equation. It is reasonable to assume a small reflection coefficient to reduce this nonlinear equation to a linear differential equation:

if $\rho^2 \ll 1$ then

$$\frac{d\rho}{du_1} = \frac{1}{2}\frac{d(\ln Z_0)}{du_1} - 2\rho = 0 \tag{1.15}$$

This differential equation can be transformed into [4]

$$\rho(\beta) = \int_0^L \frac{1}{2} \frac{d(\ln Z_0)}{dx} e^{-j2\beta x} \, dx \qquad (1.16)$$

where L is the length of the line shown in Figure 1.3, and β is the phase constant.

Several approximate solutions have been suggested as impedance taper profiles, each representing different characteristics. An exponential taper is given here as an example [4].

$$\ln(Z_0) = \frac{x}{L} \ln(Z_L) \qquad (1.17a)$$

or

$$Z_0 = e^{(x/L)\ln(Z_L)} \qquad (1.17b)$$

$$\rho(\beta) = \frac{1}{2} e^{-j\beta x} \ln(Z_L) \frac{\sin\beta L}{\beta L} \qquad (1.18)$$

which was obtained by evaluating the integral for the reflection coefficient given in equation (1.16) by assuming a constant β with respect to x. This reflection coefficient has a sin x/x behavior. Several useful impedance tapers and their corresponding reflection coefficients are given by Bolinder [6]. Defining a new function

$$P(x) = \frac{1}{2} \frac{d \ln (Z_0)}{dx} \qquad (1.19)$$

and substituting in the integral of equation (1.16) gives

$$\rho(\beta) = \int_0^L P(x) e^{-2j\beta x} \, dx \qquad (1.20)$$

Close examination of this integral reveals that the reflection coefficient is indeed the Fourier transform of $P(x)$ and that the other half of the transform is given by

$$P(x) = \frac{1}{2} \frac{d \ln (Z_0)}{dx} = \frac{1}{\pi} \int_{-\infty}^{\infty} \rho(\beta) \, e^{j2\beta x} \, d\beta \qquad (1.21)$$

This transform pair is a very useful synthesis tool. It is possible to construct a particular impedance taper governed by $P(x)$ to realize a predetermined reflection coefficient function $\rho(\beta)$. Bolinder [6] examines this synthesis problem in great depth and presents an accurate analogy with the antenna synthesis procedure that has been extensively used with great success.

Klopfenstein [8] presents an optimum length taper based on the Dolph-Chebyschev taper, which provides a minimum reflection coefficient for a given length or a minimum length to realize a given reflection coefficient. Figure 1.4 shows the reflection coefficient for such a taper as compared to an exponential taper.

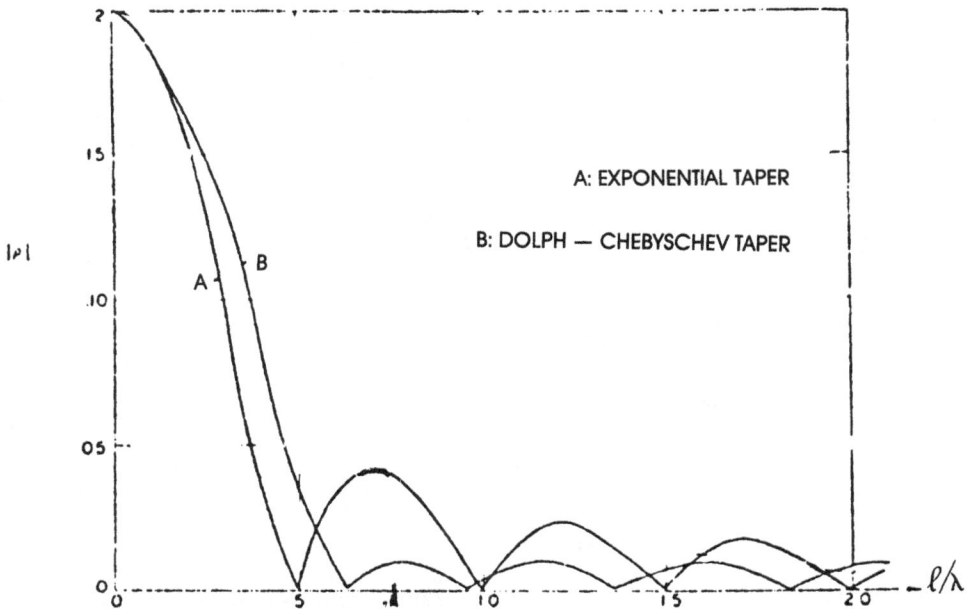

Figure 1.4 Comparison of an optimum length taper using Dolph-Chebyschev function and an exponential taper (after Klopfenstein [8], Fig. 6).

The Klopfenstein taper has an impedance jump discontinuity at two ends of the taper, which will give two impulse reflections when the proper Fourier transformation is examined. It is possible, however, to improve the taper by removing this impedance discontinuity, as done by Hecken [9]. The trade-off is a slight increase in the optimum taper length. Hecken's design, therefore, is a near optimum design. Figure 1.5 compares it with the optimum length Dolph-Chebyschev. The figure shows that, in the Hecken near optimum design, the impedance taper curve is smoothly transitioned to the impedances at both ends. The response of the near optimum taper is thus a modification of the Dolph-Chebyschev. A discussion comparing these tapers and some others is given by Hall [10] and the interested reader is encouraged to consult this article for additional information.

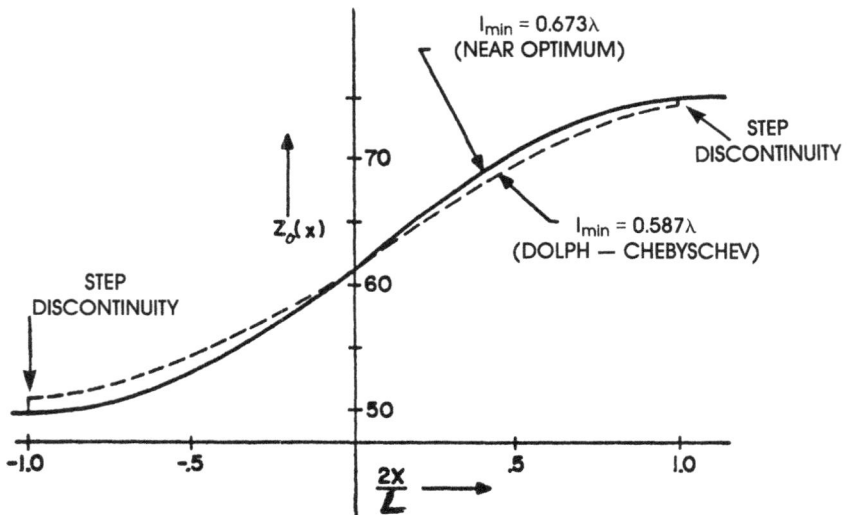

Figure 1.5 Comparison of impedance taper profile for transition from 50 Ω to 75 Ω for the optimum Dolph-Chebyschev and the near optimum due to Hecken (after Hecken [9], Fig. 5).

Often, the realization of a continuous impedance taper is difficult, and an approximation can be made by stepped transmission line tapers without too much loss of performance. An optimum design can be made by using the Dolph-Chebyschev expansion as introduced by S. Cohn [4], which gives an equal ripple response over the frequency as illustrated in Figure 1.4. A maximally flat response is obtained by using the binomial expansion. In either case, the length of the transformer sections is normally a quarter of the guide wavelength. In order to obtain improved results, the step junctions where the discontinuities occur must be compensated for the capacitance due to the step jump; the procedure and a table for proper compensation of these step discontinuities can be found in [11].

1.5 CONCLUSION

A general transition can change the propagation mechanism from one type of a media to another. This includes a simple impedance transition in like transmission lines (e.g., between coaxial lines, microstrip-lines, waveguides) or a more complicated transition from unlike media (e.g., the transition from a waveguide to microstrip, microstrip to coaxial line, waveguide to microstrip-line, or transitions to antennas). With the general transition discussion and definition in this chapter, it is possible to define some systematic steps in an attempt to provide a unified design procedure for transitions of various media. The designer can follow the following steps, using those that apply to the situation at hand:

1. Identify the two types of media to be transformed (i.e., waveguide and microstrip, *etcetera*).
2. Identify the impedance values to be matched (i.e., 500 Ω in waveguide to 50 Ω in microstrip) and the proper impedance definitions and their expressions.
3. Identify the approximate field distribution and geometry in the two media.
4. If the two line fields are drastically different, a field match must be made by designing a proper mechanism and a mechanical design. (See reference [1] for more detail.)
5. If the field match is already available go to step 6.
6. Design a proper impedance taper (steps) to achieve a transition from one impedance to the other by the procedure outlined in this book.
7. Identify the design insertion loss, the return loss, and the taper length or the number of steps (if step transition is used).
8. Identify all the design restrictions, electrical, mechanical, thermal, and so on.
9. Using a computer and the pertinent equations, develop a CAD program to aid in designing the impedance taper.
10. Provide a mechanical design and technical drawings for the machine shop.

11. Make the pertinent electrical designs and various masks for the electrical circuit fabrication.
12. Test the transition by some proper procedure (back to back or matched impedance), and compare it with the design data.
13. If design criteria are met, the design is complete, go to step 15.
14. Otherwise, re-evaluate design data and procedure, make adjustments, and reiterate the procedure until a satisfactory design is obtained.
15. Document the procedure for future use.

REFERENCES

[1] Izadian, J. "Unified Design Approach Aide Transition Design," *Microwaves and RF,* May 1987.
[2] Brews, J. R., "Characteristic Impedance of Microstrip Lines," *IEEE Trans. Microwave Theory Tech.,* Vol. MTT-35, No. 1, January 1987, pp. 30–34.
[3] Gardiol, F., *Introduction to Microwaves,* Artech House, Norwood, MA, 1984.
[4] Cohn, S., "Optimum Design of Stepped Transmission Line Transformers, *IRE Trans. Microwave Theory Tech.,* Vol. MTT-3, No. 4, April 1955, pp. 16–21.
[5] Collin, R. E., *Foundations for Microwave Engineering,* McGraw-Hill, New York, 1966.
[6] Bolinder, F., "Fourier Transforms in the Theory of Inhomogeneous Transmission Lines," *Elanders Boktryckeri Aktiebolag,* Stockholm, 1951.
[7] Ahmed, M. J., "Impedance Transformation Equations for Exponential, Cosine-Squared, and Parabolic Tapered Transmission Lines," *IEEE Trans. Microwave Theory Tech.,* Vol. MTT-29, No. 1, January 1981, pp. 67–68.
[8] Klopfenstein, R., "A Transmission Line Taper of Improved Design," *Proc. IRE,* Vol. 44, No. 1, January 1956, pp. 31–35.
[9] Hecken, T., "A Near Optimum Matching Section Without Discontinuities," *IEEE Trans. Microwave Theory Tech.,* Vol. MTT-20, No. 11, November 1972, pp. 734–739.
[10] Hall, A. H., "Impedance Matching by Tapered or Stepped Transmission Lines," *Microwave Journal,* March 1966, pp. 109–111.
[11] Matthei, G., L. Young, and E. M. T. Jones, *Microwave Filters, Impedance-Matching Networks, and Coupling Structures,* Artech House, Norwood, MA, 1980.

Chapter 2
Coaxial-to-Microstrip Transitions

2.1 INTRODUCTION

Coaxial to microstrip line transitions are very common in the *microwave integrated circuits* (MIC). Coaxial lines are used mainly to interconnect various MIC modules. Often, the microstrip circuitry of the MIC module is enclosed in a package with a coaxial connector, using a coaxial to microstrip line transition. This transition can be achieved in several ways, some of which will be presented here. Also, some design rules and the trade-offs of various approaches will be considered.

As outlined at the end of Chapter 1, the design of a transition starts with a definition of the design goals and criteria, and then proceeds through the systematic steps of actual design, realization, and evaluation of the transitions. One goal in the realization of a coaxial to microstrip line transition is the provision for a broadband impedance match and very low insertion loss. Usually mechanical and thermal consideration will have to be included in the design's requirements, so these considerations will be mentioned as points of concern when necessary and will not be elaborated on within this book.

The most widely used coaxial to microstrip transition is shown in Figure 2.1. The coaxial connector is connected to the wall of the housing through a properly designed hole or a flange. The center conductor of the coaxial connector is extended out of the connector and over the microstrip line. There are as many connectors as there are manufacturers. The main attributes to keep in mind are the frequency response of the connector to be used in the transition, mechanical considerations like size, and packaging considerations like hermeticity and thermal expansion; these must be included in the design cycle for proper operation.

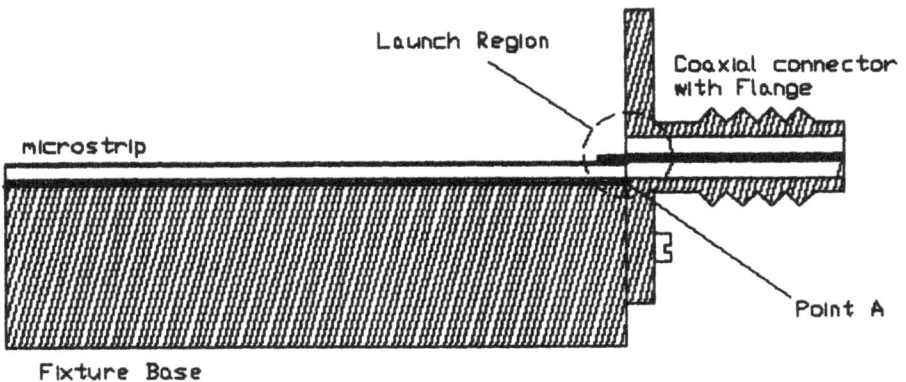

Figure 2.1 Transition from coaxial line to microstrip; the important ground
contact point, point A, is also shown.

Mode-free coaxial line connectors recently have been developed that
operate up to a 50 GHz range. This type of frequency of operation dictates
that the inner and outer diameters of the connector be very small, with
very high conductivity, and a smooth metal surface. The reason for these
requirements will shortly be explained. There are many varieties of coaxial
line connectors, designed to operate at various frequencies and for different applications. A summary of most popular ones and their operating
frequency ranges are given in Appendix A. Appendix B is a glossary of
common coaxial connector terms, included to help the reader to become
familiar with those used in practice.

Under normal operating conditions, the coaxial line is expected to
support only TEM wave propagation, however, this operation is limited
to a short range of frequencies for a given coaxial line structure, after
which the excitation of undesired higher order modes becomes possible.
In order to help visualize the excitation of undesired modes in the coaxial
line, consider the simplistic unfolded coaxial line shown in Figure 2.2a.
In a localized sense, the coaxial line may be compared to a parallel-plate
waveguide. If the upper and lower conductors are perfect conductors and
plate separation is much smaller than the cut-off wavelength of the next
higher order mode, a TEM wave will be propagating as the lowest TM
mode [1].

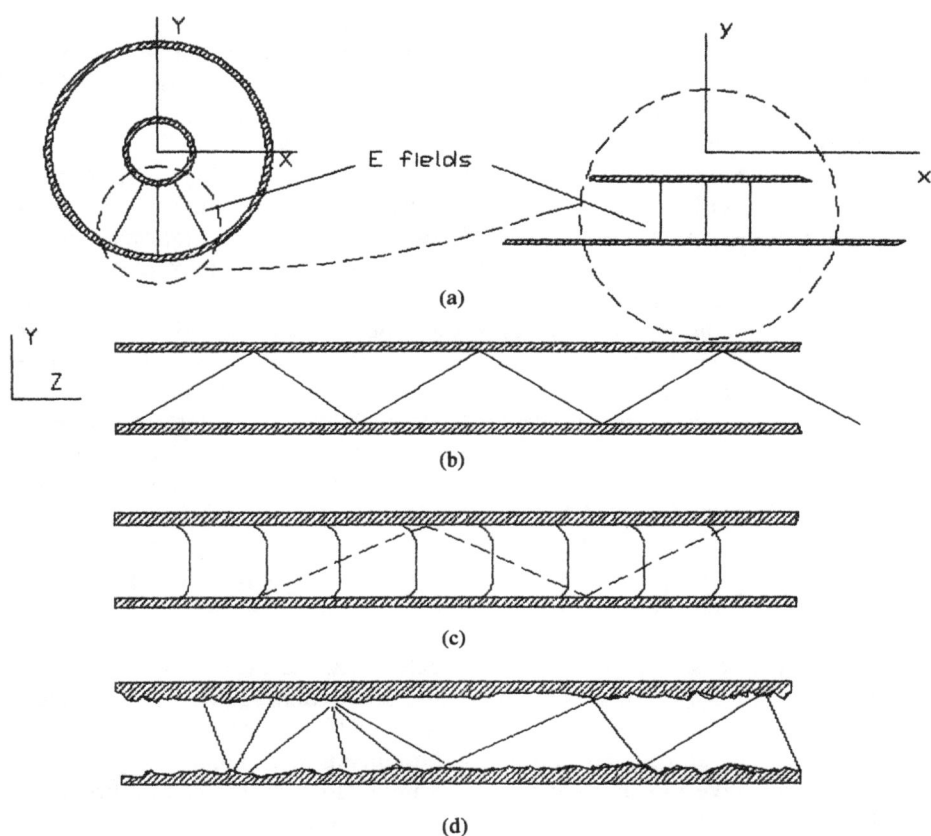

Figure 2.2 Illustration of the transverse mode excitation for coaxial lines: (a) a coaxial line unfolds to parallel plate; (b) a parallel-plate waveguide transverse resonance illustrates the localized phenomenon of resonance in coaxial lines; (c) imperfect conductors bend the fields and change direction of a pointing vector, thus potentially causing higher order modes; and (d) surface roughness also could cause higher order mode excitation.

If the plate separation is comparable to the cut-off wavelength of the next TM modes, that is, if the plate separation is half the order of multiples of the wavelength, higher order modes will propagate as illustrated in Figure 2.2b. Furthermore, if conductors are imperfect, as shown in Figure 2.2c, there will be a nonzero electric field in the conductors, forcing the field to bend and the direction of the propagation vector (pointing vector) will be tilted, thus potentially exciting the transversal propagation mode, which is undesirable. Therefore, to avoid excitation of higher order modes the separation of the two plates must be kept at less than the cut-off wavelength of the transverse mode [1]; also a very high conductivity metal, such as gold, should be utilized. Finally, one other possible source of excitation of higher order modes can be the surface roughness of the metal plates, as illustrated in Figure 2.2d.

The simplicity of the structure, then, allows easy determination of the proper dimensions for the parallel-plate waveguide. However, returning to the coaxial line configuration, the same conclusion still holds but the determination of the cut-off wavelength is much more complex and requires solving nontrivial boundary condition problems involving Bessel functions, which results in transcendental equations in terms of those functions evaluated at the radii of the inner and outer conductors.

Following the procedure outlined in Chapter 1 for the design of transitions, it is clear that the field match and the impedance match for the microstrip and the coaxial line is readily available. This is illustrated in Figure 1.3, which shows a simple way of looking at a microstrip as it is evolved by cutting and unfolding the coaxial line. The field of the two structures have the same form and an impedance match can be obtained by synthesizing a correct microstrip width and height.

In the design process, it is helpful to start by choosing a microstrip whose thickness is close to the dielectric thickness of the coaxial line and of a similar dielectric value. Furthermore, the width of the microstrip must be close to the thickness of the coaxial center pin. There may be some trade-offs, as for the electrical design, to realize the correct impedance and choose a proper substrate.

Another important issue in this transition design is the ground contact in the transition. Not only is it important to have a good coaxial center line microstrip transition, it is also essential to have good ground contact, especially directly under transition point A, as shown in Figure 2.1. This requires careful mechanical design, so that every aspect of the problem is considered.

2.2 DE-EMBEDDING AND CHARACTERIZATION OF
S-PARAMETERS

Chapter 7 presents the subject of de-embedding in detail, but it will be introduced here to ease the understanding and flow of the subject of the equivalent models, which will be discussed in the next section.

De-embedding is the process of deducing the impedance of a *device under test* (DUT) from measurements made at a distance from it. If the electrical parameters of the intervening connectors, adapters, and transitions are known. Others, such as the scattering parameters and the immitance matrices, can be deduced mathematically from the overall measured data [2].

In order to de-embed the characteristics of a device between any two transitions, it is essential to extract the S-parameters of the transition directly or develop a circuit model of the transition. Many approaches have been suggested in the literature, some of which will be examined in the coming sections. It is helpful to first consider the S-parameter characterization of the transitions.

S-parameter characterization method is suggested by Souza and Talboys [3]. This method is based on microstrip line of length $L = 2L_0 + \Delta L$ between two identical coaxial-to-microstrip connectors. Figure 2.3 shows a 50 Ω microstrip line and coaxial connectors and the corresponding signal flow graph. For the analysis to follow, it is assumed that the connector pair are symmetric and that a good degree of repeatability is achieved by replacing the connectors. Furthermore, it is assumed that there is only one discontinuity, at the junction between the coaxial connector and the microstrip line. Any of these assumptions may be checked experimentally before the start of the measurement to ensure accuracy of the transition S-parameters.

The procedure to calculate the S-parameters of the coaxial to microstrip transition is as follows:

1. Measure the transmission coefficient of a microstrip line length of $2L_0$. This measurement provides the amplitude of the transmission coefficient ($S_{12}S_{21}$) and its phase angle. The measurement is simply

$$T_0 = S_{12}S_{21} \tag{2.1}$$

where small reflections are assumed.

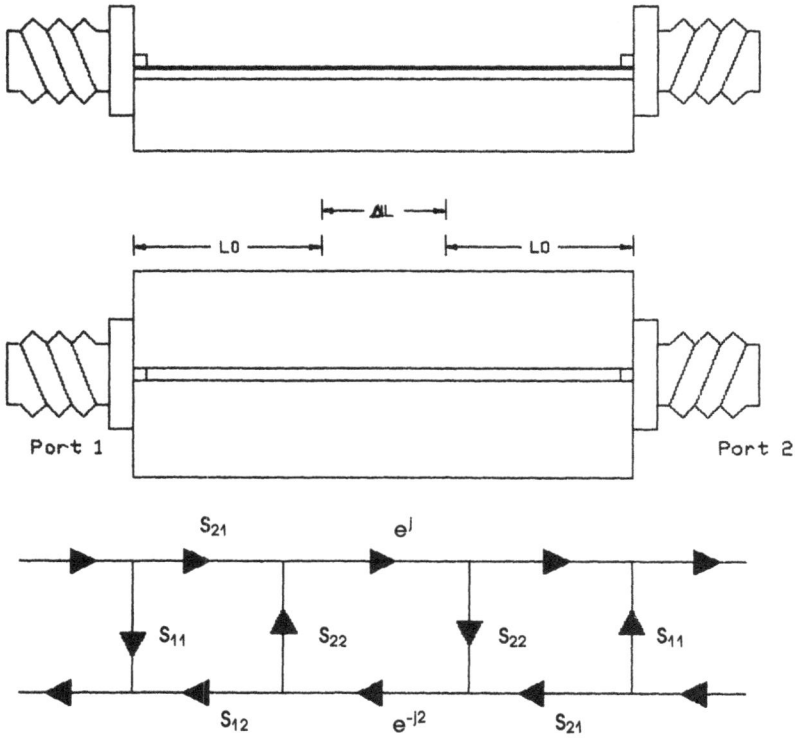

Figure 2.3 Two coaxial to microstrip transitions back-to-back and the corresponding signal flow graph.

2. Terminate one end of the fixture with a 50-Ω matched load and measure the reflection coefficient of the fixture at the other end:

$$\Gamma_{in}^0 = \frac{S_{11} - S_{22}\Delta S}{1 - S_{22}^2} \qquad (2.2)$$

where

$$\Delta S = S_{11}S_{22} - S_{12}S_{21} \qquad (2.3)$$

3. Measure the reflection coefficient, (Γ_{in}'), at one end when the other end is terminated to a sliding load of small reflection, $|\Gamma_1| < .1$. The

reflection coefficient is measured for several (at least four) phase angles of the sliding load to construct a polar plot of Γ'_{in}/Γ_1. These points are then fitted with a circle:

$$\Gamma'_{in} = \frac{S_{11} + S_{22}S_{12}S_{21} + \Gamma_1(S_{12}S_{21})^2}{1 - \Gamma_1(S_{12}S_{21}S_{22})} \tag{2.4}$$

4. Replace the $2L_0$ long microstrip with a $2L_0 + \Delta L$ microstrip and match one end of the fixture to a 50-Ω termination. The reflection coefficient of this setup is then measured:

$$\Gamma^1_{in} = \frac{S_{11} - S_{22}\Delta S \exp(-2j\theta)}{1 - S_{22}^2 \exp(-2j\theta)} \tag{2.5}$$

and

$$\theta = 2\pi\Delta L/\lambda_g \tag{2.6}$$

where λ_g is the guide wavelength.

To calculate S_{11} and S_{22}, use the equations for the two reflection coefficients found in steps 2 and 4 and the coefficient from steps 1 and 3:

$$S_{11} = j\frac{\Gamma^0_{in} \exp(-j\theta_1) - \Gamma^1_{in} \exp(j\theta_1)}{2 \sin\theta_1} \tag{2.7a}$$

$$S_{22} = j\frac{\Gamma^1_{in} - \Gamma^0_{in}}{2T_0 \sin\theta_1} \exp(j\theta_1) \tag{2.7b}$$

$$\theta_1 = \frac{2\pi (2L_0)}{\lambda_g} \tag{2.7c}$$

Special care must be exercised when finding the phase of the product term $S_{12}S_{21}$ from the reflection data. The phase ambiguity can be cleared up by comparing the calculated data and the measured data. Simplified relations for the calculation of the S-parameters of the transition are given by Suaza and Talboys [3], which are a function of frequency F (in GHz). These relations seem to have been based on results at a 2-GHz range for N-type coaxial connectors, and should be used with caution at higher frequencies and with other connectors:

$$|S_{11}| = 0.01F - 0.016 \tag{2.8a}$$

$$\arg S_{11} = -20.45F + 232.51 \tag{2.8b}$$

$$|S_{22}| = 0.031F - 0.054 \tag{2.8c}$$

$$\arg S_{22} = -84.1F + 394.59 \tag{2.8d}$$

$$|S_{12}S_{21}| = -0.047F + 1.079 \tag{2.8e}$$

$$\arg S_{12}S_{21} = -141.1F + 362.11 \tag{2.8f}$$

2.3 COAXIAL-TO-MICROSTRIP EQUIVALENT CIRCUIT

2.3.1 Element Values Deduced from Measured Return Loss

The simplest coaxial-to-microstrip transition is obtained by connecting the center pin of the coaxial connector to the microstrip trace, as shown in Figure 2.1. Wight *et al.* [4] have proposed a simple but practical experimental method for characterizing the transition. Their launcher characterization technique consists of using the launcher to make input impedance measurements on various lengths of open-circuit 50-Ω lines, up to one wavelength long, and examining the phase difference between the measured and the calculated reflection coefficients, which is referred to as the *excess phase*.

Under ideal conditions, if the launcher has no parasitic effects such as inductance and capacitance, the phase difference would remain zero or constant, as shown in Figure 2.4a. However, if the launcher exhibits any parasitic effects, the phase difference will vary accordingly. A periodic variation of the phase difference shown in Figure 2.4a suggests the presence of a shunt parasitic capacitor and a series parasitic inductor, as shown in Figure 2.4c. The reasoning behind this model is that, for the open-ended 50-Ω microstrip connected to the launcher transition, at microstrip line lengths close to odd multiples of a quarter-wavelength, the phase of the reflection coefficient seems to be dominated more by the series inductive element. This is because the open-ended microstrip line will reflect a short to ground at the point where the launcher-transition is connected to the microstrip line. For microstrip line lengths of half-wavelength orders, the phase of the reflection coefficient is dominated by the shunt capacitive element in the model. Again, this is because at this length the open-ended microstrip line will reflect an open circuit at the point of the transition.

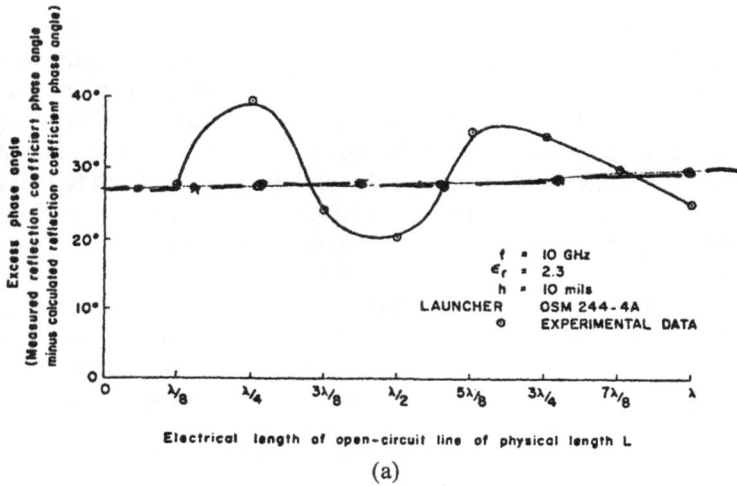

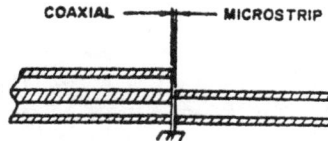

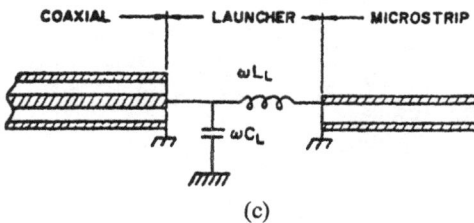

Figure 2.4 (a) The periodic variation of phase for the launcher-transition connected to an open ended 50 Ω microstrip transmission line in the presence of parasitic effects on the launcher transition as compared to one with no parasitic; (b) the equivalent circuit with an ideal transition; and (c) one with a parasitic (after Wight *et al.* [4], Fig. 2).

24

The Smith chart may be utilized to obtain the reactive values of the parasitic. It is evident that when the inductive reactance dominates the input impedance, the input impedance will be located on the upper half of the impedance chart, as shown in Figure 2.5. Thus, in order to obtain the value of the parasitic series inductor, the normalized reactance at the frequency of the measurement is read and then unnormalized to solve for the inductance. A similar approach is taken for the shunt capacitive susceptance using the admittance or the inverted Smith chart.

After the proper parasitics are accounted for in the calculated value of the reflection coefficient, the phase difference between the measured and the calculated reflection coefficient should be near zero. Any deviation of the phase difference from zero is likely to be due to additional discontinuities in the circuit that must be accounted for properly. Finally, in order to increase the accuracy of the deduced values of the parasitics, it is essential to account for the open-end effects of the microstrip by proper compensation of the line length [4].

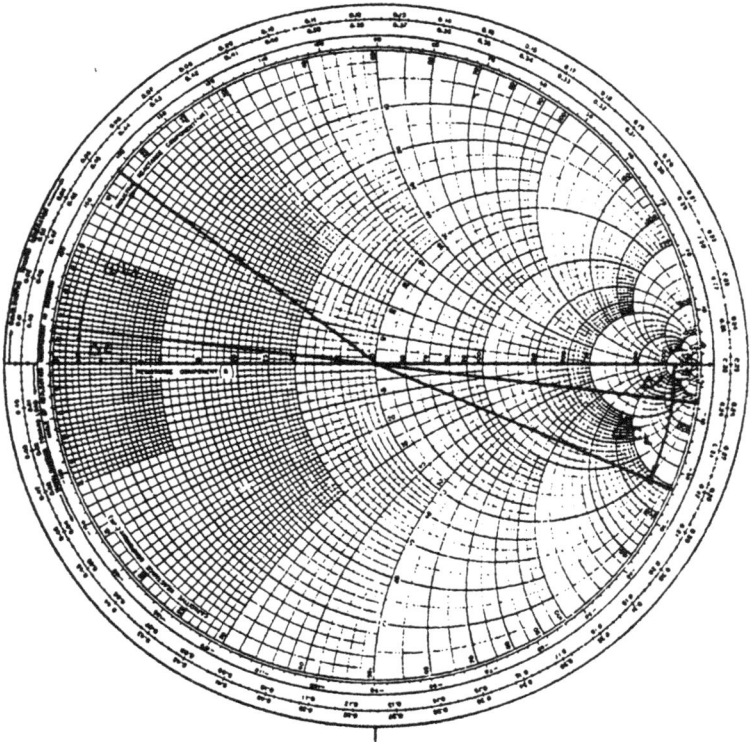

Figure 2.5 Impedance points for the parasitic elements of the transition, shown on the Smith chart (after Wight *et al.* [4], Fig. 3).

Another, similar method, reported by Gourley and Chapman [5], is based on the magnitude and phase of the return loss on open-circuited launcher-microstrip. In order to model the transition, an excess phase term was defined, as earlier, to be the difference between the measured phase and theoretical phases of the return loss, which excludes the effect of the transition parasitic. In their models, Gourley and Chapman [5] also included the effects of the dispersion as a frequency-dependent intrinsic impedance of the microstrip and the open-end effect as an incremental length of the microstrip, as shown in Figure 2.6. Figure 2.6 also shows the equivalent T-model of the transition, which is made of an incremental length of line due to launcher. Figure 2.6 shows the corresponding diagram of the test setup and the various reference planes for the circuit model. An optimization routine is used to find the component values of this model by fitting the measured data with the theoretical data. The accuracy of this model has been verified by Gourley and Chapman [5], using an experiment that included the effects of the loss in the actual measured data.

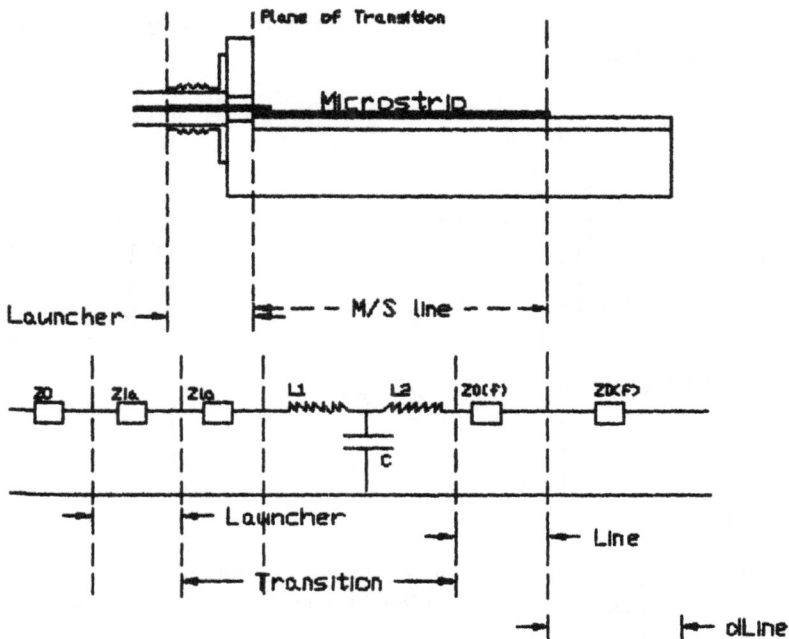

Figure 2.6 Equivalent circuit of the launcher transition after Gourley and Chapman [5] (Fig. 4). The dispersion effect of the microstrip is included as frequency-dependent intrinsic impedance, and the open-end effect is included as the incremental length of line. The T-parasitic model of the transition is also shown.

A broadband model for a coaxial-to-stripline transition is presented by Chapman and Aitchison [6]. The model parameters are deduced from the magnitude and the phase of the reflection coefficient. This method is similar to that used for the microstrip, and it will not be elaborated upon.

2.3.2 Element Values Deduced from Insertion Loss

A method for modeling the transition from a coaxial to a lossy microstrip line is presented by Majewski [7] which uses the insertion phase and insertion loss to deduce the parasitic. A π impedance model is obtained experimentally to predict the insertion loss and the insertion phase over a wide frequency range. A launcher-microstrip-launcher configuration is used to find the elements of the model in the Figure 2.7, which shows a lossy microstrip line between the parasitic models of the two-end launcher-transitions. The launchers on either side of the lossy microstrip is represented by the π-network comprising of two shunt capacitance elements connected by a series inductive element. The element values of the model are obtained from the phase and the magnitude of the insertion loss of the experimental setup.

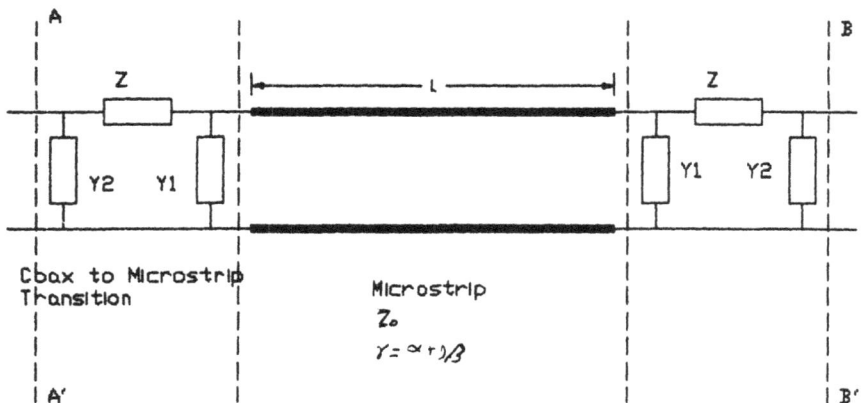

Figure 2.7 **Equivalent circuit model, characterizing the launcher-microstrip-launcher transition (after Majewski [7]) and including the lossy microstrip.**

The *insertion loss, L_I*, of the configuration of Figure 2.7 is measured at two consecutive frequencies in the band of interest. An expression describing the L_I of a load and generator terminated two-port network is given by

$$L_I = 10 \log \left| \frac{V_0}{V_1} \right|^2$$

$$= 10 \log \frac{P_a}{P_1} \, dB \qquad (2.9a)$$

$$\phi = \arg \frac{V_0}{V_1} \qquad (2.9b)$$

where P_a is the maximum power available from the generator; P_1 is the power delivered to the load Z_1 when the two-port network is inserted between the generator with impedance Z_g and the load impedance Z_1; V_0 is the voltage across Z_1 when connected directly to the generator; and V_1 is the voltage across Z_1 with the two-port inserted. It is assumed that the V_0 is at zero phase and that all $Z_1 = Z_g = Z_0$ are real impedances.

The L_I loss is then obtained from Figure 2.7 by using the chain matrix notation to cascade the input launcher model, the microstrip line, and the output launcher model to obtain the overall chain matrix:

$$\begin{bmatrix} a_{11} & a_{12} \\ a_{21} & a_{22} \end{bmatrix} = \begin{bmatrix} 1 + Y_1 Z & Z \\ Y_1 + Y_2 + Y_1 Y_2 Z & 1 + Y_2 Z \end{bmatrix}$$

$$\begin{bmatrix} \cosh\gamma 1 & Z_0 \sinh\gamma 1 \\ Y_0 \sinh\gamma 1 & \cosh\gamma 1 \end{bmatrix} \begin{bmatrix} 1 + Y_2 Z & Z \\ Y_1 + Y_2 + Y_1 Y_2 Z & 1 + Y_1 Z \end{bmatrix} \qquad (2.10)$$

where Y_1 and Y_2, Z, Z_0, and γ are defined in Figure 2.7. Furthermore, the power ratio

$$\frac{P_a}{P_1} = \left| \frac{a_{21} Z_1 Z_g + a_{22} Z_g + a_{11} Z_1 + a_{12}}{Z_g + Z_1} \right|^2 \qquad (2.11)$$

This expression can be reduced for a complete matched case, where $a_{11} = a_{22}$:

28

$$\frac{P_a}{P_1} = \left| a_{11} + \frac{a_{12}}{2Z_0} + \frac{a_{21}Z_0}{2} \right|^2 \tag{2.12}$$

also

$$\frac{P_a}{P_1} = 0.25\{[(sb_1 + ub_2)\cos\theta - (tb_2 + vb_1)\sin\theta]^2$$

$$+ [(sb_2 + ub_1)\sin\theta + (tb_1 + vb_2)\cos\theta]^2\} \tag{2.13}$$

and

$$\tan\theta = \frac{(s + um)\tan\theta + (tm + v)}{(sm + u) - (t + vm)\tan\theta} \tag{2.14}$$

$$s = (1 - y_1z)(1 - y_2z) - z(y_1 + y_2 - Y_1y_2z) \tag{2.15a}$$

$$t = z(1 - y_1z) + (1 - y_2z)(y_1 + y_2 - y_1y_2z) \tag{2.15b}$$

$$u = 0.5[(1 - y_1z)^2 + (1 + y_2z)^2 - (y_1 + y_2 - y_1y_2z)^2 - z^2] \tag{2.15c}$$

$$v = (1 - y_1z)(y_1 + y_2 - y_1y_2z) + z(1 - y_2z) \tag{2.15d}$$

where

$$b_1 = 2\cosh(\alpha 1) \tag{2.16a}$$

$$b_2 = 2\sinh(\alpha 1) \tag{2.16b}$$

$$m = \coth(\alpha 1) \tag{2.16c}$$

$$z = Z/Z_0 \tag{2.16d}$$

$$y_{1,2} = Y_{1,2}/Y_0 \tag{2.16e}$$

$$\theta = \beta 1 = 2\pi \sqrt{\epsilon_{\text{eff}}(f)}/\lambda_0 \tag{2.16f}$$

and $\epsilon_{\text{eff}}(f)$ is the dispersive effective dielectric constant.

For the simplified case Y_1 can be neglected and after some mathematical simplifications and manipulations; the values of the parasitic elements become

$$C = \frac{-b \pm \sqrt{b^2 - 4ac}}{2a} \tag{2.17a}$$

$$b = Z_0 = [\xi(f_2) - \xi(f_1)] \tag{2.17b}$$

$$a = Z_0^2 [\psi(f_2) - \psi(f_1)] \tag{2.17c}$$

$$c = \eta(f_2) - \eta(f_1) \tag{2.17d}$$

and

$$L = \frac{1}{C} (\eta(f_1) - Z_0 C \xi(f_1) - Z_0^2 C^2 \xi(f_1)) \tag{2.17e}$$

where

$$\eta(f_i) = \frac{(m + 1)(k_i - \tan\theta_i)}{\omega_i^2[(2m + 1)k_i - (m + 2) \tan\theta_i]} \tag{2.18a}$$

$$\xi(f_i) = \frac{(m + 1)(1 + k_i \tan\theta_i)}{\omega_i[(2m + 1)k_i - (m + 2) \tan\theta_i]} \tag{2.18b}$$

$$\psi(f_i) = \frac{0.5(k_i - m \tan\theta_i)}{[(2m + 1)k_i - (m + 2) \tan\theta_i]} \tag{2.18c}$$

$k_i = \tan\phi_i$, where ϕ_i is measured phase at frequency i.

$$\omega_i = 2\pi f_i$$

The phase ϕ is periodic with frequency f, thus it is possible to take two values $f_2 > f_1$ so that $\phi = 0 \pm \pi/2$ to simplify these expressions [7].

To obtain a more broadband model, Y_1 can also be included. The broadband expressions for the element value are thus given by

$$L = \frac{1}{(C_1 + C_2)} [\eta(f_1) - Z_0(C_1 + C_2)\xi(f_1)$$
$$- Z_0^2(C_1 + C_2)^2 \psi(f_1)] \tag{2.19}$$

where

$$(C_1 + C_2) = \frac{-b \pm \sqrt{b^2 - 4ac}}{2a} \qquad (2.20)$$

Having calculated these inductance and capacitances, a fit to the measure data is made by optimizing the ratio of the two capacitances. Always keep in mind that this method assumes symmetry.

These measurements are used to verify the accuracy of the correlation between the measured and the theoretical values of these models, as reported in [7]. Figure 2.8 shows partial results of these measurements for a standard SMA connector and a 50-Ω microstrip transmission line. In Figure 2.8, the two frequencies, f_1 and f_2, are shown at two consecutive zero phase points at 8.09 and 10.75 GHz, respectively. Figure 2.9 shows a theoretically calculated insertion phase and loss as a function of frequency. It is apparent that when the parasitics of the transition are set to zero (the ideal case), the calculated and the measured phases differ significantly. It is also observed in Figure 2.9 that the insertion phase is relatively insensitive to the microstrip loss and, therefore, so are the values of the parasitic elements. The insertion loss, as might be expected, varies with the microstrip loss factor, as shown in Figure 2.9. In this analysis, the SMA launchers are assumed to be lossless. However, it is possible to include the effect of this loss in the model by including a resistive network here [7].

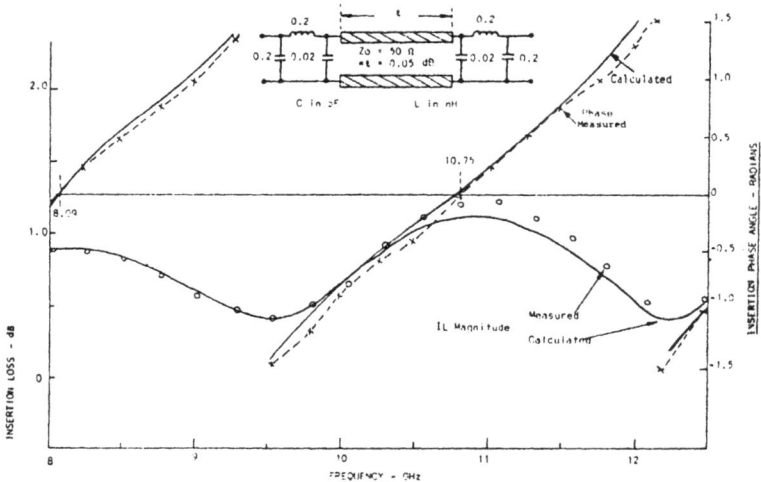

Figure 2.8 Comparison of the calculated and measured data for the equivalent circuit of Figure 2.7 (after Majewski [7], Fig. 5).

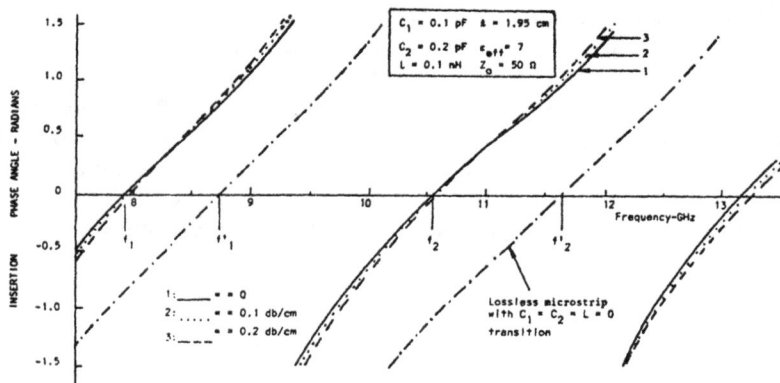

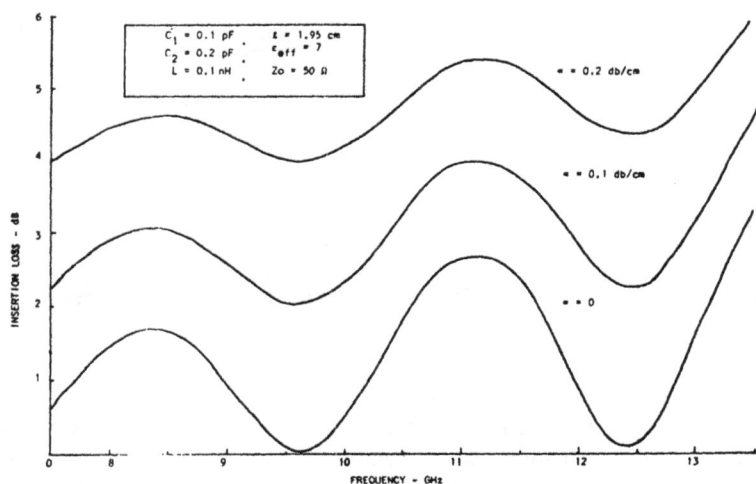

Figure 2.9 Calculated insertion loss and the phase *versus* frequency (after Majewski) [7], Fig. 7).

2.4 OTHER USEFUL TRANSITIONS

Often, the need for a broadband transition arises that requires a special consideration because it cannot be realized by the conventional off-the-shelf launchers. Special designs are possible that provide wider band transitions of coaxial-to-microstrip lines.

A quarter-wave stepped transition is often used to make a transition from coaxial to microstrip line. In designing this type of a transition, a predetermined reflection can be realized in a minimum number of steps. Each step is realized by a quarter-wave-long impedance transformer, a typical step transition is shown in Figure 2.10. Proper compensation for the step discontinuities at each end of the transformer line needs to be made for fringing capacitances by reducing the transformer length slightly to account for the fringing fields at the step plane [8] in addition to a short inductive (high impedance) line section to resonate the discontinuity capacitance. An excellent discussion and graphical representation of such transition can be found in [9].

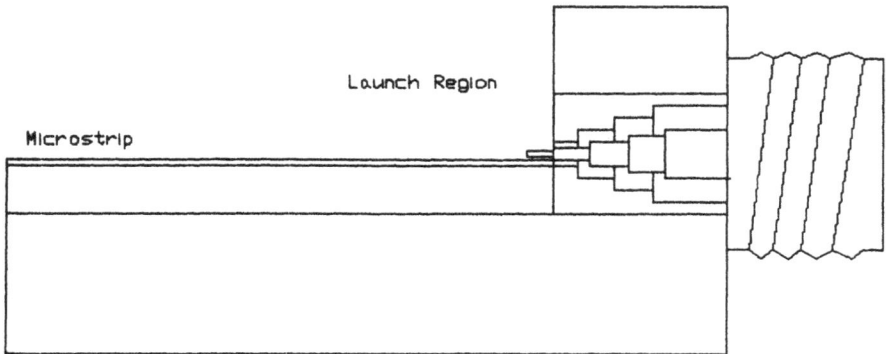

Figure 2.10 **Realization of a step coaxial line transition to achieve a wide band transition.**

In addition to the quarter-wave-long stepped transformers, a continuous taper may be used as shown in Figure 2.11. The figure shows a coaxial line tapering from larger to smaller diameter lines that will provide a field geometry similar to that of the microstrip line. This kind of taper may be considered the limit of the stepped transition process, as the number of the steps approaches infinity and the length of each transformer section tends toward zero.

Eisenhart [10] has described a useful design, shown in Figure 2.12. It uses a rotationally asymmetric coaxial taper. This configuration provides a tapered transition from the coaxial to an off-center (eccentric) coaxial line. The electric field line pattern at the off-centered end of the taper looks similar to that of the microstrip line, which satisfies the field match condition for making an efficient transition, as outlined at the end of Chapter 1. The impedance match is achieved by adjusting the line's inner radius and the offset of the center conductor in the taper gradually along the taper length. A design curve is provided by Eisenhart [10]. This

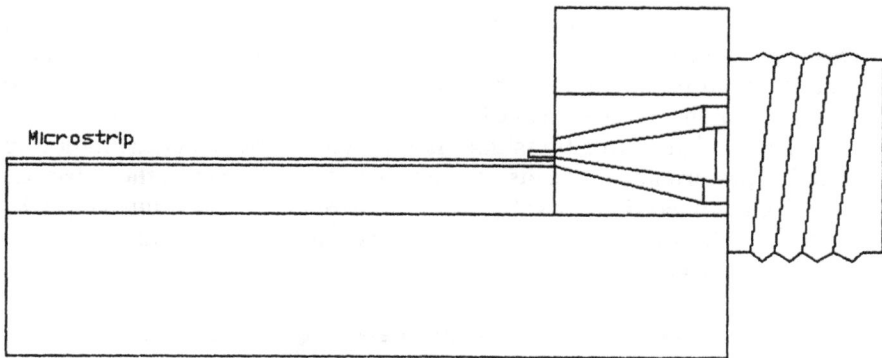

Figure 2.11 A continuous taper coaxial line; both the inner and outer diameters are tapered.

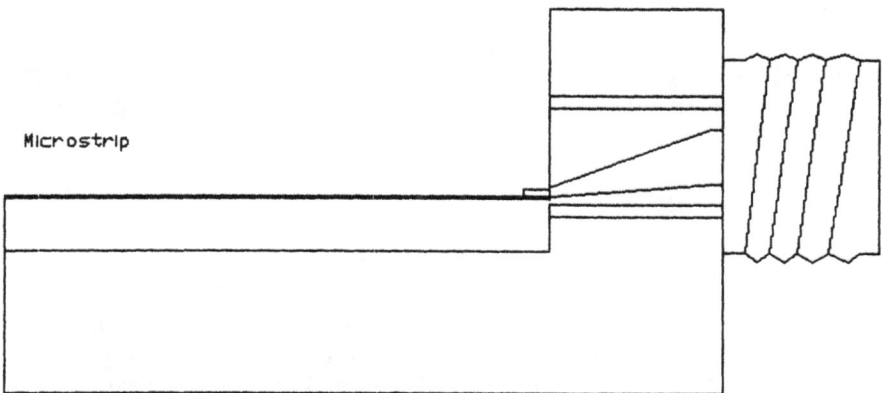

Figure 2.12 A continuous taper transition (taper after Eisenhart [10]); at the end of the transition, the field geometry is similar to that of the microstrip line.

curve, however, does not lend itself to a computer-aided design. A rigorous analysis of the offset center-conductor coaxial line is needed to obtain the impedance of the line as a function of the line's physical dimensions. A complete analysis of the transmission line structure is presented in [11], which will be very helpful in the design. Also an approximate solution to various irregular transmission lines is given by Pan [12], which may be used to model the eccentric line.

34

Finally, a right-angle coaxial-to-microstrip transition may be desired. This is possible by drilling a hole in the microstrip substrate, inserting the connector pin here, and soldering the shield of the coaxial connector to the ground of the microstrip line [13]. This configuration is shown in Figure 2.13. The design of this transition may require some experimentation to optimize performance, because the impedance of the pin in the substrate is not well defined. It is possible to use *time domain reflectometry* (TDR) to identify the discontinuities and incorporate proper compensation to optimize the transition [14, 15].

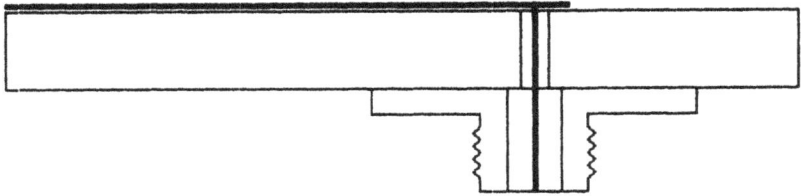

Figure 2.13 A right-angle transition made by inserting the coaxial connector perpendicular to the bottom of the substrate and connecting the center connector to the microstrip trace.

REFERENCES

[1] Shen, L. C., and J. A. Kong, "Applied Electromagnetism," Brooks/Cole Engineering Division, Monterey, CA, 1983, p. 94.
[2] Bauer, R. F., and P. Penfield, "De-embedding and Unterminating," *IEEE Trans. Microwave Theory Tech.*, Vol. MTT-22, No. 3, March 1974.
[3] Souza, J. R., and E. C. Talboys, "S-Parameter Characterization of Coaxial to Microstrip Transition," *IEE Proc.*, Vol. 129, Part H, No. 1, February 1982.
[4] Wight, J. S., O. P. Jian, W. J. Chudoblak, and V. Makios, "Equivalent Circuit of Microstrip Discontinuities and Launchers," *IEEE Trans. Microwave Theory Tech.*, Vol. MTT-22, No. 1, January 1974.
[5] Gourley, S. E., and A. G. Chapman, "Broadband Characterization of Coaxial to Microstrip Transitions," 12th European Microwave Symposium, 1982.

[6] Chapman, A. G., and C. S. Aitchinson, "A Broad-Band Model for a Coaxial-to-Stripline Transition," *IEEE Trans. Microwave Theory Tech.*, Vol. MTT-28, No. 2, February 1980.

[7] Majewski, M. L., R. W. Rose, and J. R. Scott, "Modeling and Characterization of Microstrip-to-Coaxial Transitions," *IEEE Trans. Microwave Theory Tech.*, Vol. MTT-29, No. 8, August 1981.

[8] Matthie, G., L. Young, and E. M. T. Jones, *Microwave Filters, Impedance-Matching Networks, and Coupling Structures*, Artech House, Norwood, MA, 1980.

[9] Hoffmann, R. K., *Integrierte Mikrowellenschaltungen*, Springer-Verlag, Heidelberg, 1983 (translation: H. Howe, ed., *Handbook of Microwave Integrated Circuits*, Artech House, 1987).

[10] Eisenhart, R. L., "A Better Microstrip Connector," IEEE International Microwave Theory and Techniques Symposium Digest, 1978, p. 318.

[11] Hilberg, W., *Electrical Characteristics of Transmission Lines*, Artech House, Norwood, MA, 1979.

[12] Pan, S., "Approximate Determination of the Characteristic Impedance of the Coaxial System Consisting of an Irregular Outer Conductor and a Circular Inner Conductor," *IEEE Trans. Microwave Theory Tech.*, Vol. MTT-35, No. 1, January 1987.

[13] Hammerstad, E. O., and Bekkadal, F., "Microstrip Handbook," University of Trondheim, Norwegian Institute of Technology (unpublished manuscript, no date).

[14] Made-It Associates, "Precision Right-Angle Coaxial to Microstrip Lines," *Mama's Notes*, Vol. 6, No. 1, January–February 1987.

[15] Made-It Associates, "Analysis of Right Angle Coaxial-to-Stripline Transition," *Mama's Notes*, Vol. 6, No. 6, November–December 1987.

Chapter 3
Waveguide-to-Coaxial Line Transitions

3.1 INTRODUCTION

The rectangular waveguide is a non-TEM wave propagating structure, where the wave is transmitted in a TE or TM mode. The dominant mode of operation for the rectangular waveguide is the TE_{10}. The guide wavelength, the cut-off frequencies, and the guide impedance for the TE_{10} mode are well known. The useful bandwidth of operation in the rectangular waveguides is determined by the cut-off frequency of the TE_{10} and that of the next higher mode, TE_{20}. This bandwidth usually starts at 25 percent over the cut-off frequency of the TE_{10} mode and ends at the cut-off frequency of the TE_{20} [1]. Thus, the rectangular waveguides may be thought of as having a high pass–band-pass property.

In contrast to rectangular waveguides, a coaxial line supports a TEM mode and, thus, is very broadband as compared to the rectangular waveguides. Furthermore, the coaxial lines can be used from dc to the upper frequency limit of the operation, where the transmission loss becomes significant or the higher modes are excited, whichever occurs first (see the discussion in Chapter 2).

A well designed transition from coaxial line-to-rectangular waveguide thus will have, at best, the bandwidth of the waveguide. Primarily, two classic approaches have been taken in designing coaxial line-to-waveguide transitions: the electric probe and the magnetic probe. The major portion of the design concerns finding the best location, height, and diameter of the probe to achieve an optimum impedance match. The impedance of the TE_{10} mode in a rectangular waveguide is of the order of few hundred Ohms, whereas the coaxial line usually will have the much lower impedance of few tens of Ohms, typically 50 Ω. This requires a high transformation ratio for the impedance match to the probe. There are, however, some techniques to optimize the impedance match, and these will be presented shortly.

3.2 ELECTRIC PROBE WAVEGUIDE LAUNCHERS

Cohn [2] presents a special class of waveguide-to-microstrip connector that connects the inner conductor of a coaxial line to the top side of the waveguide wall, and the shield of the coaxial line is connected to the bottom side as shown in Figure 3.1b. Cohn uses the "voltage-current" definition to characterize the impedance of the TE_{10} mode of the waveguide, as discussed in Chapter 1.

$$Z_0 = 600/\sqrt{\epsilon} \cdot b/a\sqrt{1 - (f_c/f)^2} \tag{3.1}$$

$$\lambda_g = \lambda/\sqrt{1 - (f_c/f)^2} \tag{3.2}$$

In making a transition from the 50-Ω coaxial line to the high impedance of the waveguide, it is possible to construct a reduced dimension waveguide that will exhibit a 50-Ω impedance based on the impedance formula (3.1) just given. This step is the first critical transition between the coaxial line and a waveguide of unlike field geometry but the same impedance, as outlined in Chapter 1.

As illustrated in Figure 3.1, the final stage of the transition design (i.e., transition from a 50-Ω waveguide to higher impedance of the normal-size waveguide) can be achieved by one or several quarter-wave waveguide sections or a continuous impedance taper, as outlined in Chapter 1. An approximate equivalent circuit for this transition is given in Figure 3.1c. As illustrated in Figure 3.1b and 3.1a, the center conductor extends into the waveguide and terminates on the top waveguide wall. It is well known that this will look like an inductive post in the guide [3], thus causing mismatches. To reduce this inductance and provide a better match, the center coaxial pin is widened gradually, as it extends into the waveguide, as shown in the Figure 3.1b. The optimum diameter of the post is reported to be .15a (where a = the guide width [2, 3]). A good starting position for the post is a quarter-wavelength from the shorted end, centered in the waveguide. This quarter-wave line will act as an impedance transformer. As for any quarter-wave impedance transformer, the bandwidth is limited to the regions at which the quarter-wave length is not violated too drastically. Ultimately, the bandwidth is limited at the lower end by the cut-off frequency of the guide and on the high side by the excitation of the TE_{20}. It is believed that the coaxial line post will not excite the TE_{20} because it connects to the point of zero electric field in the waveguide (if that can be located practically). Excitation of higher order modes in the guide is possible but can be avoided with proper compensation, the reader is invited to consult [2] for additional ideas on broadening the frequency response of the guide.

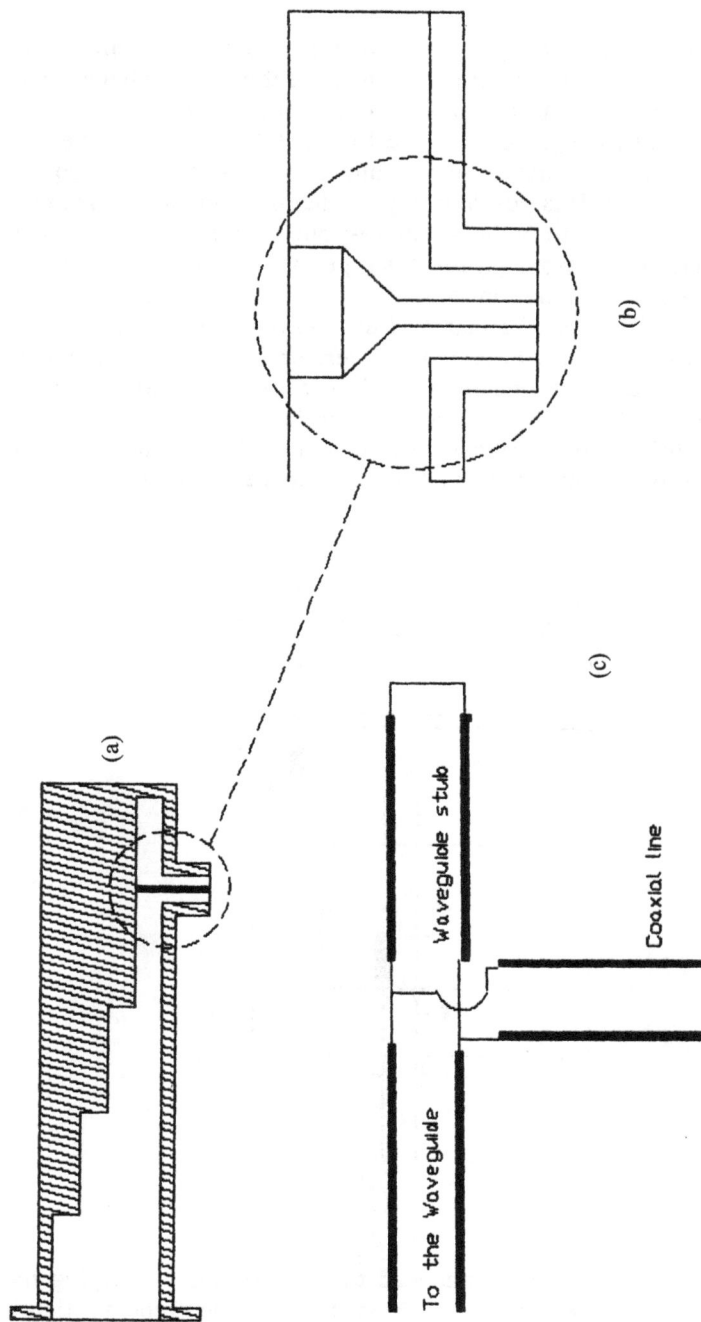

Figure 3.1 (a) Reduced-height 50-Ω waveguide and coaxial line, where the center conductor is extended into the waveguide to form the probe; (b) widened coaxial center conductor to compensate inductance; and (c) equivalent circuit for the waveguide transitions.

40

There are many variations of this transition that principally work in the same way. The most common variation of this approach has the center pin of the coaxial line just extending into the waveguide and not connecting to the opposite waveguide wall (see Figure 3.2). This resembles a small monopole antenna radiating in the guide, thus exciting the propagation mode in the guide. It is possible to vary the location and the geometry of the monopole probe to optimize the response. An optimization approach is presented by Mumford [4] for this type of a transition, also, a detailed moment method formulation for such a problem is reported by Jarem [5].

A coaxial line can be matched to a waveguide via a probe antenna located ahead of a short-circuited plunger. An impedance match can usually be achieved by varying any two of the three dimensional parameters: probe length, off-center position of the probe, and piston position in the waveguide [4]. Figure 3.2 illustrates the monopole antenna probe and the parameters to be optimized to obtain an impedance match.

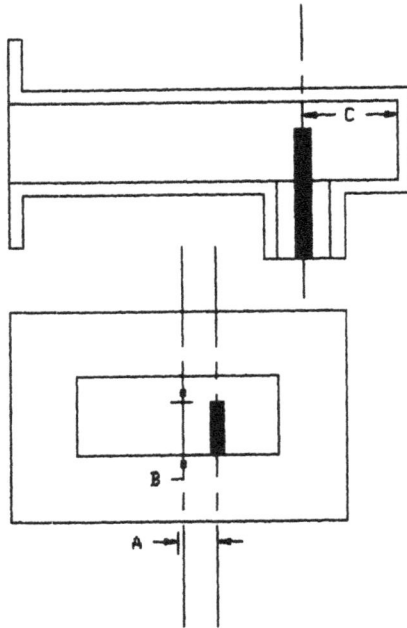

Figure 3.2 A probe may be matched to a waveguide by varying the off-center position, the probe length, and the probe position.

The waveguide impedance changes with frequency, the reactance of the probe also varies with frequency, which makes the broadband impedance match more difficult. However, the following procedure will provide an optimum design. For a short probe that is connected to the opposite wall of the waveguide, the impedance can be expressed as [4]

$$Z = Z_{0g} \sin^2 \frac{2\pi L}{\lambda_g} \cos^2 \frac{\pi d}{a} + jX = R + jX \qquad (3.3)$$

where L is the length from shorting piston to probe (C in Figure 3.2); λ_g is the wavelength in the guide; d is the distance of the probe from the center line of the guide (A in Figure 3.2); a is the width of the guide; X includes reactive components introduced by the probe inductance and the shorting piston; and Z_{0g} is the characteristic impedance of the waveguide.

The real part of this equation can be matched to the impedance of the coaxial line. It is seen that by changing the piston position, L, or the off-center position, d, any resistance less than the characteristic impedance of the guide can be obtained. But it may change rapidly with frequency if the wrong piston position is used; consequently, the band might be unnecessarily narrow.

The bandwidth of the transducer depends partly upon how rapidly this resistive component departs from matching value as the frequency is changed, and will tend to be greatest when the slope of R (the real value of the impedance) is zero with respect to the wavelength (i.e., when R is maximum). This would occur with the piston a quarter-wavelength from the probe if the characteristic impedance of the waveguide was constant. Ordinarily, however, this guide impedance changes with the wavelength, so that the maximum resistance occurs when the piston position is somewhat less than a quarter-wavelength.

The piston position that yields an unchanging value of R, and hence the greatest bandwidth for an excellent impedance match, may be derived from real part of (3.3) by setting the first derivative of R to zero with respect to guide wavelength. This gives the following transcendental equation:

$$\frac{\tan (2\pi L/\lambda_g)}{2\pi L/\lambda_g} = 2/(\lambda/2a)^2 \qquad (3.4)$$

So far in these calculations, the reactive mismatch has been neglected, and it must be matched next. In the foregoing discussions, we assumed that the current in the probe was uniform, which is correct only for guides with b-dimensions (heights) that are very small compared to the guide wavelength and when the probe is connected directly to the opposite side wall of the guide. In most practical guides, however, the waveguide is so high and the probe is so long that the current cannot be assumed to be uniform, and an impedance transformation takes place along the probe. This impedance transformation may be accounted for by considering the probe as an antenna and obtaining a coupling coefficient. Such a coupling coefficient can be assigned to a probe antenna whether it extends all the way across the guide and connects to the other side or merely projects part of the way into the waveguide and is left open. The open-circuited probe is of practical importance in wideband transducers, since its controllable capacitance can be used to resonate out the inductive reactance introduced by the piston, which is located at the optimum position to help widen the operating frequency range of the transition. Mumford [4] provides a systematic, straightforward design procedure that can be very helpful, and the interested reader is encouraged to refer to this classic paper.

An elaborate, extensive discussion of the probe-fed waveguide is given by Collin [6]. In this approach, the probe is treated as monopole antenna radiating in the short-circuited waveguide. The Green function for this situation is used to find the radiated power and the stored energies around the probe. The radiation resistance, which is also the real part of the probe input impedance, is found from the total radiated power; and the reactance part of the input impedance of the probe is found from the stored energies around the probe. The final results from this analysis is given here for convenience, but the details of the derivation are omitted for brevity. The input impedance of the probe antenna, as seen from the coaxial line, is

$$Z_{in} = R + jX \tag{3.5a}$$

$$R = \frac{2Z_0}{ab\beta_{10}k_0} \sin^2 \beta_{10}l \, \tan^2 k_0 \frac{d}{2} \tag{3.5b}$$

$$X = \frac{Z_0 \tan^2 k_0 \frac{d}{2}}{2\pi k_0 b} \left\{ \ln \frac{2a}{\pi r} + \frac{0.0518 k_0^2 a^2}{\pi^2} + \frac{2\pi}{\beta_{10}a} \sin 2\beta_{10}l \right.$$
$$\left. - 2 \left(1 - \frac{2r}{a} \right) - 2k_0^2 \sum_{m=1}^{\infty} \left[1 - \frac{\sin^2 (m\pi d/2b)}{\sin^2 (k_0 d/2)} \right]^2 \frac{K_0(k_m r)}{k_m^2} \right\} \tag{3.5c}$$

$$k_m^2 = (m\pi/b)^2 - k_0^2 \tag{3.5d}$$

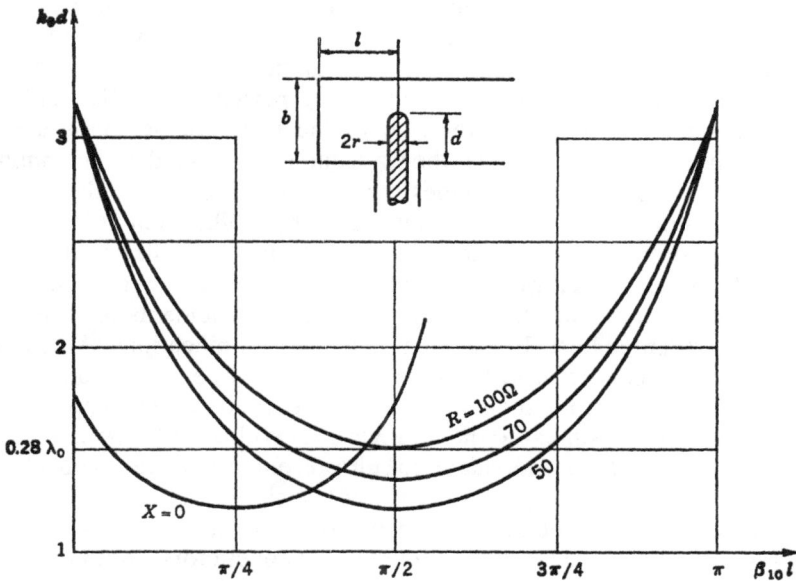

Figure 3.3 A graphical solution for obtaining the optimum dimensions for a probe transition after Collin [6].

where Z_0 is the free-space impedance, $\beta_{10} = (\pi/a)^2 - k_0^2$, and dimensions are as shown in Figure 3.3. K_0 is the modified Bessel function of the second kind.

These equations can be used to plot the contours of the constant R and $X = 0$ as functions of the physical location and the shape parameters of the probe. The intersection between the zero reactance contours and the contours of any resistance values will provide the optimum design parameters. Figure 3.3 illustrates the simplicity of such a graphical approach for the particular case shown. In this figure, the intersections of the $X = 0$ with contours of the constant R's to establish the optimum dimensions for the probe transitions.

Eisenhart et al. [7] presented an analysis of a coaxial-waveguide junction based on an equivalence to a waveguide post approach that Eisenhart and Khan had presented in an earlier paper [8]. Bialkowski and Khan [9] and [10] employed a modal analysis approach for the junction between a coaxial line and a generalized waveguide with upper and lower surfaces; their input admittance expression can be used to match the coaxial line to the waveguide.

A more general analysis of the waveguide-coaxial line junction is reported by Williamson [11]. Figure 3.4 shows three typical structures that can be modeled by this technique. The junction between the coaxial line and the waveguide is treated as a "two-gap" junction for which the admittance matrix is found. In this formulation, it is assumed that the waveguide and coaxial line are conducting perfectly and that only the TEM mode can propagate in the coaxial line at the frequency of interest. Furthermore, it is assumed that the waveguide is operated at the dominant TE_{10} mode. The analysis provides the admittance matrix and the corresponding equivalent circuit. A later report by Williamson [12] analyzed a configuration similar to that in Figure 3.4a but with dissimilar coaxial lines, the outer coaxial radii were different, as shown in Figure 3.4c. The equivalent circuits and the corresponding circuit elements for this latter case are presented here. It is evident that small modifications of the model and geometry can be used to adjust the coaxial line-to-waveguide transition; for example, by short circuiting the second gap in Figure 3.4b, a transition junction similar to that of Figure 3.1 is obtained. For this reason, the equations and the equivalent circuit presented here can be useful for analyzing or synthesizing these types of transitions.

The element of the admittance matrix for the junction of Figure 3.4c is given for the two dissimilar coaxial outer conductors, b_1 and b_2:

$$Y_{11} = -\frac{2\pi j}{\eta_0 kh \ln^2(b_1/a)} \left\{ kh \ln (b_1/a) \cot (kh) - D_0^{11} - 2 \sum_{m=1}^{\infty} D_m^{11} \right\}$$

$$(3.6)$$

$Y_{22} = Y_{11}$ by replacing b_1 for b_2.

$Y_{21} = Y_{12}$ by reciprocity

$$= -\frac{2\pi j}{\eta_0 kh \ln (b_1/a) \ln (b_2/a)}$$

$$\cdot \left[\ln (b_1/a) \frac{kh}{\sin (kh)} - D_0^{21} - \sum_{m=1}^{\infty} (-1)^m D_m^{21} \right] \qquad (3.7)$$

$$D_m^{11} = -\frac{\pi}{2}(Y_0(q_m'ka)J_0(q_m'kb_1) - Y_0(q_m'kb_1)J_0(q_m'ka)$$

$$\cdot \left(\frac{J_0(q_m'kb_1)}{J_0(q_m'ka)} + j \frac{(J_0(q_m'kb_1)Y_0(q_m'ka) - J_0(q_m'ka)Y_0(q_m'kb_1))}{J_0(q_m'kb_1)S^*(q_m'kb_1,q_m'kd_1,e/d)} \right) \Big/ q_m'^2$$

$$(3.8)$$

for $m\pi / kh < 1$.

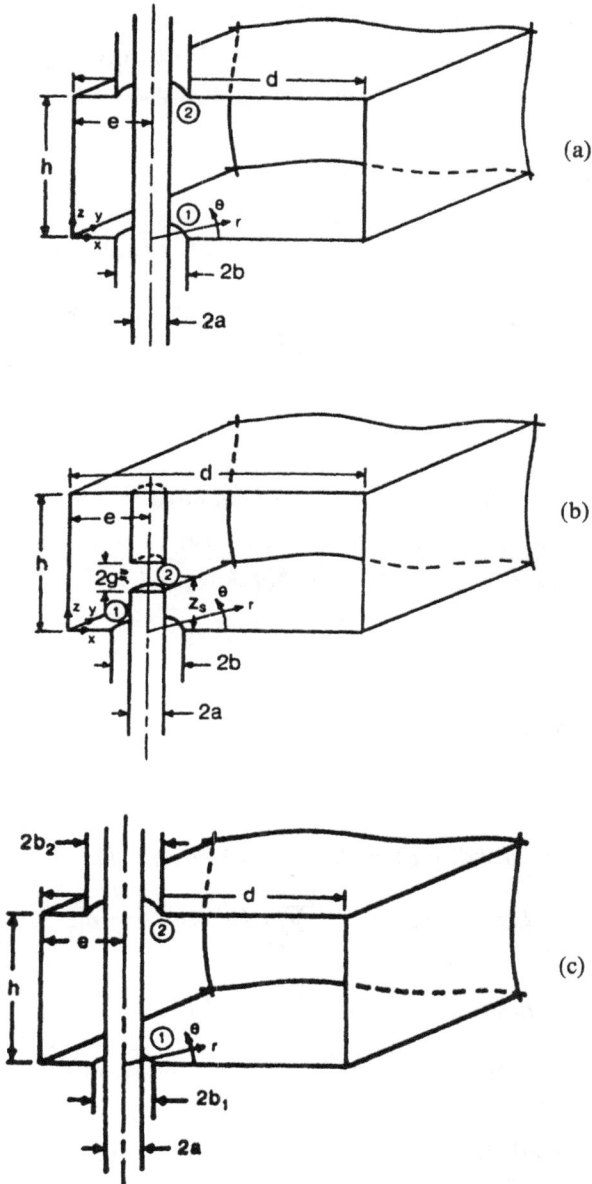

Figure 3.4 (a) The cross-coupled coaxial line–rectangular waveguide junction (b) a two-gap mounting structure and (c) dissimilar coaxial lines (after Williamson [10], Figs. 1, 2; [11], Fig. 1).

$$D_m^{11} = [K_0(q_m k b_1)I_0(q_m k a) - K_0(q_m k a)I_0(q_m k b_1)]$$

$$\cdot \left(\frac{I_0(q_m k b_1)}{I_0(q_m k a)} - \frac{(I_0(q_m k b_1)K_0(q_m k a) - I_0(q_m k a)K_0(q_m k b_1))}{I_0(q_m k a)S(q_m k a, q_m k d, e/d)} \right) \bigg/ q_m'^2$$

(3.9)

for $m\pi/kh > 1$.

$$D_m^{21} = -\frac{\pi}{2} (Y_0(q_m' k a)J_0(q_m' k b_1) - Y_0(q_m' k b_1)J_0(q_m' k a)$$

$$\cdot \left(\frac{J_0(q_m' k b_2)}{J_0(q_m' k a)} + j\frac{(J_0(q_m' k b_2)Y_0(q_m' k a) - J_0(q_m' k a)Y_0(q_m' k b_2))}{J_0(q_m' k a)S^*(q_m' k a, q_m' k d, e/d)} \right) \bigg/ q_m'^2$$

(3.10)

for $m\pi/kh < 1$.

$$D_m^{21} = [(K_0(q_m k b_1)I_0(q_m k a) - K_0(q_m k a)I_0(q_m k b_1)]$$

$$\cdot \left[\frac{I_0(q_m k b_2)}{I_0(q_m k a)} - \frac{(I_0(q_m k b_2)K_0(q_m k a) - I_0(q_m k a)K_0(q_m k b_2))}{I_0(q_m k a)S(q_m k a, q_m k d, e/d)} \right] \bigg/ q_m^2$$

(3.11)

for $m\pi/kh > 1$, and

$$q_m = \sqrt{(m\pi/kh)^2 - 1} \tag{3.12a}$$

$$q_m' = \sqrt{1 - (m\pi/kh)^2} \tag{3.12b}$$

$$S^*(q_m' k a, \ q_m' k d, \ e/d) = H_0^{(2)}(q_m' k a) + J_0(q_m' k a)$$

$$\cdot \left[\sum_{n-\infty \neq 0}^{\infty} H_0^{(2)}(2|n|q_m' k d) - \sum_{n-\infty \neq 0}^{\infty} H_0^{(2)}(2|n + e/d|q_m' k d) \right] \tag{3.13a}$$

$$S(q_m k a, \ q_m k d, \ e/d) = K_0(q_m k a) + I_0(q_m k a)$$

$$\cdot \left[\sum_{n-\infty \neq 0}^{\infty} K_0(2|n|q_m' k d) - \sum_{n-\infty \neq 0}^{\infty} K_0(2|n + e/d|q_m k d) \right] \tag{3.13b}$$

An equivalent circuit for this transition is shown in Figure 3.5; the various elements of the equivalent circuit are as follows:

$$B_a = B_{11} - B_{12} \tag{3.14a}$$

$$B_b = B_{21} \tag{3.14b}$$

$$B_c = B_{22} - B_{21} \tag{3.14c}$$

$$X_A = xZ_w + X_B/2 \tag{3.14d}$$

$$Z_W = \frac{2kh}{k_{10}d}\,\eta_0 \tag{3.14e}$$

$$k_{10}d = \sqrt{(kd)^2 - \pi^2} \tag{3.14f}$$

$$x = \frac{k_{10}d}{4\pi}\csc^2(\pi e/d)$$
$$\cdot\left\{ \ln\left(\frac{Ckd}{\pi}\sin(\pi e/d)\right) - 2\sin^2(\pi e/d) + 2\sum_{m=2}^{\infty}\sin^2(n\pi e/d)\right.$$
$$\left.\cdot\left[\frac{1}{\sqrt{n^2 - (kd/\pi)^2}} - \frac{1}{n}\right] - \frac{\pi}{2}\frac{Y_0(ka)}{J_0(ka)}\right\} \tag{3.15}$$

with $C = 1.78107. \ldots$

$$X_B = 2\pi k_{10}d(a/d)^2\sin^2(\pi e/d)Zw \tag{3.16}$$

$$B_{21} = -\frac{2\pi}{\eta_0\ln(b_1/a)\sin(kh)} + \frac{2\pi}{\eta_0 kh\ln(b_1/a)\ln(b_2/a)}$$
$$\cdot\left\{ 2\sum_{m=1}^{\infty}(-1)^m D_m^{21} + \frac{\pi}{2}\frac{J_0(kb_1)}{J_0(ka)}\right.$$
$$\left.\cdot[J_0(ka)Y_0(kb_2) - J_0(kb_2)Y_0(ka)]\right\} \tag{3.17}$$

$$B_{11} = -\frac{2\pi}{\eta_0\ln(b_1/a)}\cot(kh) + \frac{2\pi}{\eta_0 kh\ln^2(b_1/a)}$$
$$\cdot\left\{ 2\sum_{m=1}^{\infty}D_m^{11} + \frac{\pi J_0(kb_1)}{2\,J_0(ka)}\cdot[J_0(ka)Y_0(kb_1) - J_0(kb_1)Y_0(ka)]\right\} \tag{3.18}$$

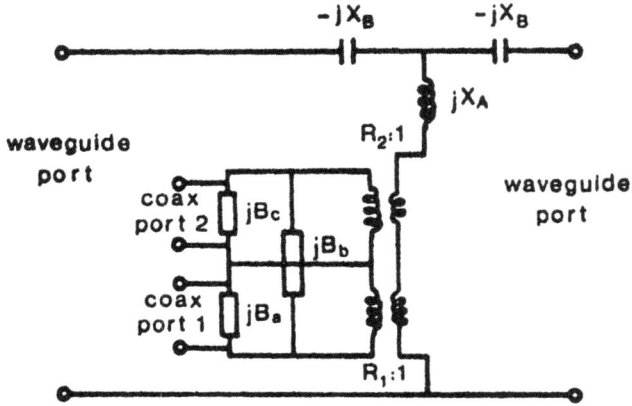

Figure 3.5 Equivalent circuit for the coaxial-waveguide junction shown in Figure 3.4 and for single mode propagation in the waveguide.

$B_{22} = B_{11}$ when b_1 is replaced by b_2 and D_m^{11} is replaced by D_m^{22}. The transfer ratios are given by

$$R_1 = \frac{(2/\pi) \ln (b_1/a) J_0(ka) \sin (\pi e/d)}{J_0(ka) Y_0(kb_1) - J_0(kb_1) Y_0(ka)} \tag{3.19a}$$

$$R_2 = \frac{(2/\pi) \ln (b_2/a) J_0(ka) \sin (\pi e/d)}{J_0(ka) Y_0(kb_2) - J_0(kb_2) Y_0(ka)} \tag{3.19b}$$

It should be noted that setting the outside diameter of any of the coaxial line equal to the inside diameter provides the typical transition of the center conductor of the coaxial line extending inside the guide and connecting to the opposite wall, as was shown in Figure 3.1. Setting the gap 2 of the Figure 3.4b to the top of the waveguide will model the probe feed waveguide as presented by Collin [6]. (See also Figures 3.2 and 3.3.) Williamson [13] has shown that these formulations provide a more accurate representation of the probe feed waveguide than those expressions provided by Collin, which were presented earlier. Thus, Collin's formulas although easier to use because they involve mostly sine functions, must be used with caution.

Additional information about these types of waveguide launchers can be found in Lewin [17].

3.3 MAGNETIC LOOP WAVEGUIDE LAUNCHERS

Another way of coupling a coaxial line to a TE_{10} mode waveguide is to use a magnetic probe antenna, which can be made by extending the center conductor of a coaxial line connected to the shorting wall of a waveguide into the guide, making a L-shape loop, and terminating the center line to the bottom broadwall of the waveguide, as shown in Figure 3.6. Several authors have studied this type of end-launcher [6, 14, 15]. This type of launcher can be analyzed by assuming some currents on the horizontal and vertical sections of the loop and self-reacting these assumed currents over the extent of the loop probe. The input impedance to the loop at the plane of the interface between the loop and the coaxial line is found from the stationary expression [16] that follows:

$$Z_{in} = - \int_v \frac{\mathbf{E} \cdot \mathbf{J}}{I_{in}^2} \, dv \qquad (3.20)$$

where $\mathbf{E}$ is the electric field inside the guide due to current $\mathbf{J}$ flowing in volume v. I_{in} is the total input current at the reference point.

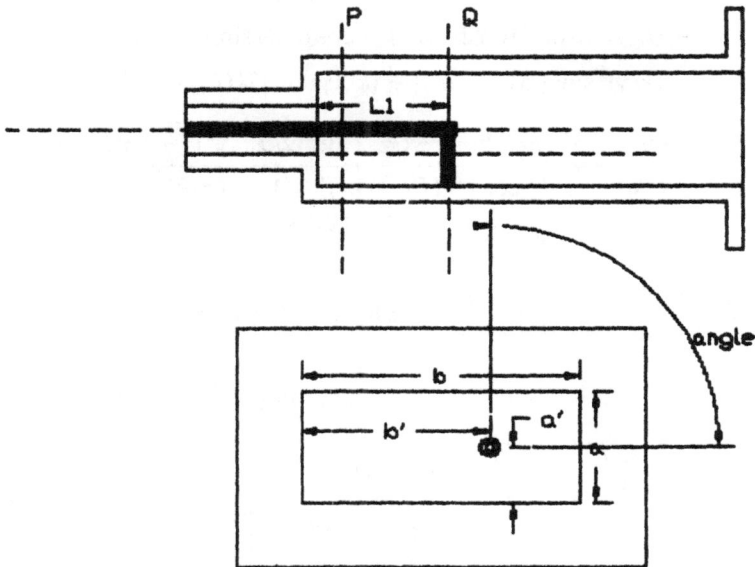

Figure 3.6 A magnetic loop for coupling to a rectangular or circular waveguide (after Dan and Sanyal [15], Fig. 1).

Omitting the mathematical details of the derivation for the input impedance, we simply provide the final expression for the input impedance and invite the interested reader to refer to [16] for detailed derivations. The input impedance has the following form:

$$Z_{in} = R_{in} + jX_{in} \tag{3.21}$$

$$X_{in} = X_1 + X_2 + X_3 + X_4 \tag{3.22a}$$

$$R_{in} = \frac{240\pi}{kakb} \frac{\sin^2(\pi b'/b) \sin^2(ka') \sin^2(kL_1 \sqrt{1 - (\pi/kb)^2})}{\cos^2[k(L_1 + a')] \sqrt{1 - (\pi/kb)^2}}$$
$$\cdot [J_0(\pi R/b)J_0(kR \sqrt{1 - (\pi/kb)^2}) \tag{3.22b}$$

$$X_1 = \sum_{n=1} \sum_{m=1} \frac{120\pi}{kakb \cos^2[k(L_1 + a')]}$$
$$\cdot F \frac{\sqrt{(n\pi/ka)^2 + (m\pi/kb)^2 - 1}}{(n\pi/ka)^2 + (m\pi/kb)^2}$$
$$[\sin^2(n\pi a'/a) \sin^2(m\pi b'/b)J_0(n\pi R/a)J_0(m\pi R/b)] \tag{3.23a}$$

$$F = e^{-\gamma L_1}[(k/\gamma) \sin(ka') + \cos(ka')]\{2(k/\gamma) \sin([k(L_1 + a')]$$
$$- (k/\gamma) \sin(ka') \cosh(\gamma L_1) + \cos(ka') \sinh(\gamma L_1)\}$$
$$+ (k/\gamma)^2 \sin^2[k(L_1 + a')] + (k/2\gamma) \sin[2k(L_1 + a')] \tag{3.23b}$$

$$X_2 = \frac{120\pi}{kakb} \frac{\sin^2(\pi b'/b) \sin^2(ka') \sin[2kL_1 \sqrt{1 - (\pi/kb)^2}]}{\cos^2[k(L_1 + a')] \sqrt{1 - (\pi/kb)^2}}$$
$$\cdot \{J_0(\pi R/b)J_0[kR \sqrt{1 - (\pi/kb)^2}]\} \tag{3.24}$$

$$X_3 = \frac{120\pi}{ka} \frac{\sin^2(ka')}{\cos^2[k(L_1 + a')]} \{.5 \ln(2b/\pi R) \sin(\pi b'/b)$$
$$- J_0(\pi R/b) \sin^2(\pi b'/b) + \frac{K^2 b^2}{2\pi^2} \cos^2(kb'/b)$$
$$- \frac{K^2 b^2}{4\pi^2} \left[\frac{2\pi b'/b)^2}{2} \ln(2\pi b'/b) \right.$$
$$\left. \left. - .75(2\pi b'/b)^2 - (2\pi b'/b)^4/288. \ldots \right] \right\} \tag{3.25}$$

$$X_4 = \frac{120}{ka \cos^2(k(L_1 + a'))} \sum_{n=1}^{\infty}$$

$$\cdot \frac{[(n\pi/ka)\sin(n\pi a'/a)\cos(ka') - \cos(n\pi a/a)\sin(ka')]^2}{(n\pi/ka)^2 - 1}$$

$$\cdot [K_0(k_n R) - K_0(2k_n b')] \tag{3.26}$$

$$k_n = \sqrt{(n\pi/ka)^2 - 1} \tag{3.27}$$

$$\gamma = \sqrt{(n\pi/a)^2 + (n\pi/b)^2 - k^2} \tag{3.28}$$

It must be noticed that, in the expressions for the input impedance, X_2 is the reactance due to the dominant mode, X_3 and X_4 are the reactances due to modes with indexes $n = 0$, $m > 1$, and $n > 0$, $m \geq 1$, respectively. These expressions can be used to develop a series of design curves by solving them as a function of the physical parameters of the transition for the locus of points, where the reactance is zero and the resistance is equal to 50 Ω.

REFERENCES

[1] Gardiol, F., *Introduction to Microwaves,* Artech House, Norwood, MA, 1984.

[2] Cohn, S. B., "Waveguide-to-Coaxial-Line Junctions," *Proc. IRE,* September 1947, pp. 920–926.

[3] Marcuvitz, N., *Waveguide Handbook,* McGraw-Hill, New York, 1951.

[4] Mumford, W. W., "The Optimum Piston Position for Wide-Band Coaxial-to-Waveguide Transducers," *Proc. IRE,* February 1953.

[5] Jarem, J. M., "A Multifilament Method of Moments Solution for the Input Impedance of a Probe-Excited Semi-Infinite Waveguide," *IEEE Trans. Microwave Theory Tech.,* Vol. MTT-35, No. 1, January 1987.

[6] Collin, R., *Field Theory of the Guided Waves,* McGraw-Hill, New York, 1960.

[7] Eisenhart, R. L., P. T. Greiling, L. K. Roberts, and R. S. Robertson, "A Useful Equivalence for a Coaxial-Waveguide Junction," *IEEE Trans. Microwave Theory Tech.,* Vol. MTT-26, No. 3, March 1978.

[8] Eisenhart, R. L., and P. J. Khan, "Theoretical and Experimental Analysis of a Waveguide Mounting Structure," *IEEE Trans. Microwave Theory Tech.,* Vol. MTT-19, No. 8, August 1971.

[9] Bialkowski, M. E., and P. J. Khan, "Modal Analysis of a Coaxial-Line Waveguide Junction," *IEEE MTT-S Int. Microwave Symp. Digest,* 1983, N-6.

[10] Bialkowski, M. E., and P. J. Khan, "Determination of Admittance of Coaxial-Line/Rectangular-Waveguide Junctions," *Electron. Lett., 29 September 1983, Vol. 19, No. 20.*

[11] Williamson, A. G., "Analysis and Modeling of 'Two-Gap' Coaxial Line Rectangular Waveguide Junctions," *IEEE Trans. Microwave Theory Tech.,* Vol. MTT-31, No. 3, March 1983.

[12] Williamson, A. G., "Cross-Coupled Coaxial-Line/Rectangular-Waveguide Junction," *IEEE Trans. Microwave Theory Tech.,* Vol. MTT-33, No. 3, March 1985.

[13] Williamson, A. G., "Design of Coaxial-Line/Rectangular-Waveguide Transitions," *Int. J. Electron.,* Vol. 58, No. 3, 1985, pp. 425–429.

[14] Harrington, R., *Time Harmonic Electromagnetics,* McGraw-Hill, New York, 1961.

[15] Dan, B. N., and G. S. Sanyal, "Coaxial-Line to Waveguide End Launchers," *Proc. IEE,* Vol. 123, No. 10, October 1976.

[16] Deshpande, M. D., B. N. Das, and G. S. Sanyal, "Analysis of an End Launcher for an X-Band Rectangular Waveguide," *IEEE Trans. Microwave Theory Tech.,* Vol. MTT-27, No. 8, August 1979.

[17] Lewin, L., *Theory of Waveguides,* Halsted Press, New York, 1975.

Chapter 4
Waveguide-to-Microstrip and Fin-Line Transitions

4.1 INTRODUCTION

An efficient transition from waveguide to microstrip requires a field match; after that, an impedance match can be achieved simply. The systematic steps to achieve a field match were outlined in Chapter 1. The fields of a TE_{10} mode in the waveguide and that of the quasi-TEM in the microstrip are shown in Figure 4.1. It is apparent from the figure that the two fields are polarized in the same way; however, the fields in the waveguide span a wide region in the cross section, whereas the fields in the microstrip are much more concentrated, located solely between the strip and the ground plane. It is thus obvious that, in order to make an efficient transition, the fields of the waveguide must be forced to concentrate in an area relatively the size of the active area in the microstrip. In this chapter, we will consider several configurations of the microstrip transition that deal with this technique. Furthermore, the forthcoming discussions will illustrate how the impedance match can be achieved in a more straightforward manner.

4.2 RIDGED WAVEGUIDE APPROACH

A ridged waveguide offers many advantages over the conventional rectangular waveguide; namely, broader bandwidth, more flexibility in the impedances, and concentration of fields in the smaller region in the waveguide. These characteristics make the ridged waveguide a perfect candidate for facilitating transition from rectangular waveguides to microstrip lines. Figure 4.1 compares a typical ridged waveguide to a conventional

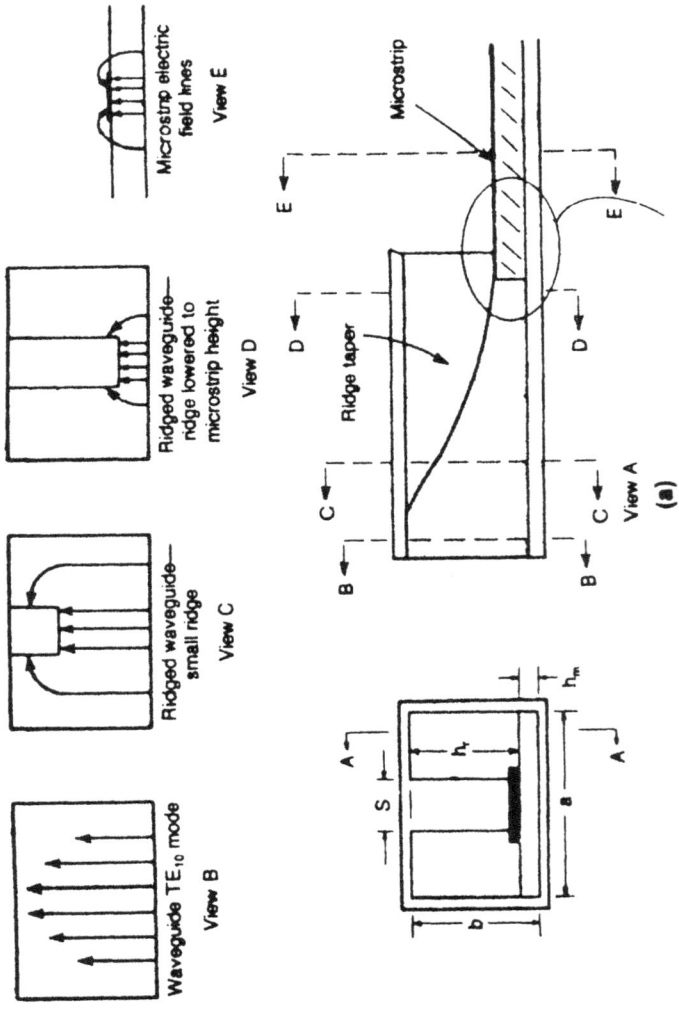

Figure 4.1. Similarities between the fields of a TE_{10} mode in the waveguide and that of the microstrip. The fields of the waveguide span the wider cross section of the guide, whereas the fields of the microstrip are concentrated between the strip and ground. A ridge waveguide can force the fields to concentrate in a smaller region under the ridge.

rectangular waveguide. A double or single ridged waveguide can be designed with the proper physical dimensions to provide almost any impedance value between the tips of the two ridges in the case of the double ridges or the tip of the ridge and the ground for those with a single ridge case. Figure 4.1 also compares the fields of the microstrip line to those of the ridged guide. It is thus apparent that a field match in the ridged guide is readily achievable [1], which turns our focus to the impedance properties of ridge waveguide, in order to achieve an impedance match to the microstrip.

The cutoff wavelength for the ridged waveguide can be obtained by simple transverse resonance techniques as outlined by Cohn [2], Hopfer [3], Pyle [4], and others. Here we provide a transcendental equation, the solution of which will provide the cut-off wavelength for the ridge guide. An extensive explanation of this method and the approach for obtaining the cut-off wavelength of the ridged waveguide is found in [4]. Marcuwitz [5] gives the relationship for the normalized shunt susceptance due to the step discontinuity seen at the side of the ridge that appears in the transcendental equation and that is illustrated in Figure 4.2:

$$1 - \frac{Z_2}{Z_1} \tan\phi_1 \tan\phi_2 - B Z_2 \tan\phi_2 = 0 \tag{4.1}$$

where

$$\phi_1 = \frac{s\pi}{\lambda_c} \tag{4.2a}$$

$$\phi_2 = \frac{(a - s)\pi}{\lambda_c} \tag{4.2b}$$

$$\frac{Z_2}{Z_1} = b/d = 1/\alpha = 1/(1 - \delta) \tag{4.2c}$$

$$
\begin{aligned}
B Z_2 = P(2b/\lambda_c) \Bigg\{ &\ln \left[\left(\frac{1 - \alpha^2}{4\alpha} \right) \left(\frac{1 + \alpha}{1 - \alpha} \right)^{(\alpha + 1/\alpha)/2} \right] \\
&+ 2 \frac{A + A' + 2C}{AA' - C^2} \\
&+ \left(\frac{b}{4\lambda_c} \right)^2 \left(\frac{1 - \alpha}{1 + \alpha} \right)^{4\alpha} \left(\frac{5\alpha^2 - 1}{1 - \alpha^2} + \frac{4\alpha^2 C}{3A} \right)^2 \Bigg\},
\end{aligned}
\tag{4.3}
$$

for $b/\lambda_c < 1$.

$$A = \left(\frac{1 - \alpha}{1 + \alpha}\right)^{2\alpha} \frac{1 + \sqrt{1 - (b/\lambda_c)^2}}{1 - \sqrt{1 - (b/\lambda_c)^2}} - \frac{1 - 3\alpha^2}{1 - \alpha^2} \qquad (4.4)$$

$$A' = \left(\frac{1 + \alpha}{1 - \alpha}\right)^{2/\alpha} \frac{1 + \sqrt{1 - (d/\lambda_c)^2}}{1 - \sqrt{1 - (d/\lambda_c)^2}} + \frac{3 + \alpha^2}{1 - \alpha^2} \qquad (4.5)$$

$$C = \left(\frac{4\alpha}{1 - \alpha^2}\right)^2 \qquad (4.6)$$

$$P = \coth\left[\frac{\pi(a - s)}{2b}\right] \qquad (4.7)$$

all of these formulas can be converted to the single ridge case by replacing λ_c with $\lambda_c/2$.

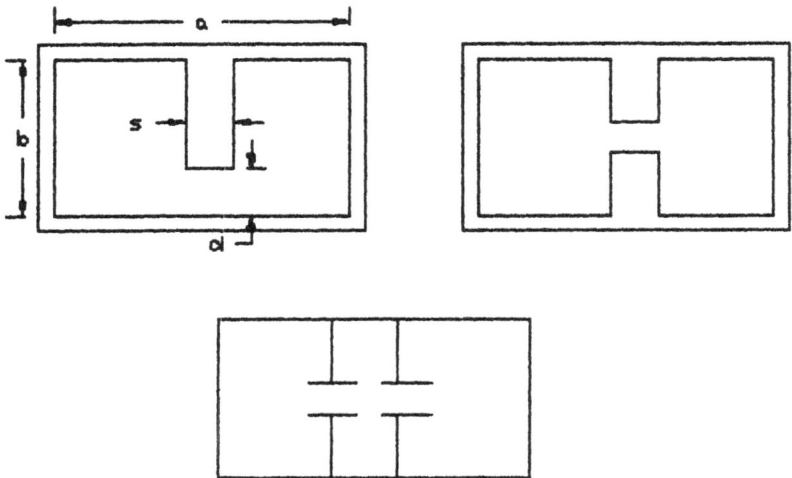

Figure 4.2. Parameters of the ridged waveguide and the equivalent circuit including a shunt capacitance (after Hopfer [3] and Pyle [4]).

To find the impedance of the ridge guide as function of the physical dimensions and the cut-off wavelength, use the following two relations. The first formula, given by Cohn [2], is based on the voltage-current definition and is good for very thin ridges, although it can be used for thicker ridges as well. The second formula, given by Hopfer [3], is based on the voltage-power definition as introduced in Chapter 1. For simplicity, admittances at the infinite frequency are given without loss of generality:

$$Y_{0\infty} = \frac{\lambda_c}{d} \frac{\sin\phi_1 + (d/b)\cos\phi_1 \tan(\phi_2/2)}{120\pi^2} \tag{4.8}$$

where $b/a \leq 0.5$, $s/a \leq 0.4$ is the implied range.

$$Y_{0\infty} = \sqrt{\frac{\epsilon_0}{\mu_0}} \frac{\lambda_c}{\pi d} \left\{ \frac{2dR}{\lambda_c} \cos^2\phi_1 BZ_2 + \phi_1/2 + 0.25\sin(2\phi_1) \right.$$

$$\left. + (d/b) \frac{\cos^2\phi_1}{\sin^2\phi_2} [\phi_2/2 - 0.25\sin(2\phi_2)] \right\} \tag{4.9}$$

In these cases the admittances are those at infinite frequency. The impedance at the design frequency can be found by

$$Z_0 = \frac{1}{Y_0} = \frac{1}{Y_{0\infty}\sqrt{1 - (\lambda/\lambda_c)^2}} \tag{4.10}$$

The cut-off wavelength of the guide is obtained by solution of a transcendental equation (4.1), a closed form expression for the cut-off wavelength is also possible, and this is given by Hoefer and Burton [6].
For very thin ridges (fins):

$$\frac{b}{\lambda_c} = \frac{b}{2a} \left[1 + (1 + 0.2\sqrt{b/a}) \frac{4b}{\pi a} \ln \csc \frac{\pi d}{2b} \right]^{-1/2} \tag{4.11}$$

For the cut-off wavelength of thick ridges this equation takes the following form:

$$\frac{b}{\lambda_c} = \frac{b}{2(a-s)} \left[1 + (1 + 0.2\sqrt{b/(a-s)}) \frac{4b}{\pi(a-s)} \ln \csc \frac{\pi d}{2b} \right.$$

$$\left. + (2.45 + 0.2s/a) \frac{sb}{d(a-s)} \right]^{-1/2} \tag{4.12}$$

where $0.01 < d/b < 1$, $0 < b/a < 1$, and $0 < s/a < 0.45$ [6].
A more general expression for the admittance is given by Hoefer and Burton [6], having the restrictions discussed earlier:

$$Y_{0\infty} = \frac{(b/d)\sin\phi_1 + [(2b/\lambda_c)\ln\csc(\pi d/2b) + \tan(\phi_2/2)]\cos\phi_1}{120\pi^2(b/\lambda_c)}$$

$$\tag{4.13}$$

A more numerical, intensive method of analysis of the ridge wave-guide is presented by Utsumi [7], who uses a variational method to obtain electric field profiles for the TE and TM modes and numerical values for the cut-off wavelength and the characteristic impedance of the ridge guide. In addition to academic interest in this method, the paper [7] can be employed to verify empirical expressions, thus allowing additional con-fidence and insight in their results.

Knowing the cut-off wavelength and the impedance characteristic to the ridge waveguide, it is now possible to design a ridge waveguide taper that will obtain a impedance of 50 Ω to match to the microstrip. The procedure to obtain a continuous taper is similar to that outlined in Chapter 1. It is also possible to design a step transition of quarter-wavelength sections, in which case proper compensation must be made for the discontinuity capacitances. To design the transition, we implement an analysis program based on the previous formulas. To realize a given impedance, a zero search routine can be implemented through an analysis program. This method can be used to synthesize the ridge guide of any impedance. It is thus very simple to design the taper profile. References [8–10] can aid in this design and also provide design examples.

4.3 WAVEGUIDE-TO-FIN-LINE APPROACH

Fin-line transmission lines are widely used at the millimeter frequency range. Often, a transition from one impedance to another is used. Such an impedance match requires adjusting the fin gaps to realize a particular impedance taper, either continuous or stepped transition. In the latter case, an equivalent model for the fin line can be made from a cascade of sections of uniform transmission lines. A continuous taper can be modeled similarly, the only requirement is the knowledge of the fin-line propagation constants for the fundamental modes.

A simple multisection reflection model can be used to analyze any such transitions. This procedure is outlined in [11]. The dielectric material on which the fin line is constructed usually causes reflections at the in-sertion points in the waveguide, where the waveguide seems to be di-electrically loaded. Even if the transition itself was reflectionless, the reflection from the front edge of the fin-line dielectric will cause reflec-tions.

The fin-line dielectric reflection can be reduced by proper compen-sation and impedance matching, as illustrated in Figure 4.3. A quarter-wave transition transformer may be designed using either a notch or protrusion cut in the fin-line substrate at the waveguide-to-fin-line inter-face. Experiments have shown improvement of at least 5 dB on the return loss of such a transition over the entire waveguide band.

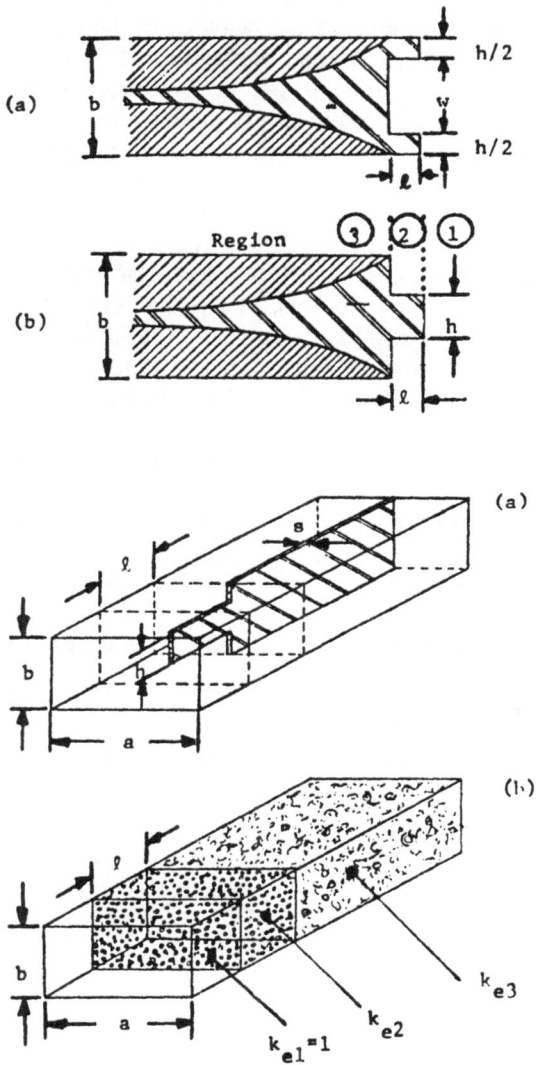

Figure 4.3. Modeling of transformer sections: (a) real structure; (b) equivalent homogeneous waveguide model (after Verner and Hoefer [12], Fig. 2).

For the impedance taper on the fin-line substrate, it is clear that, when the fin gap is as open as the waveguide height, the waveguide is essentially a dielectrically loaded waveguide. From an electromagnetic point of view, it is clear that proper precautions are needed to prevent reflections at the interface of the empty guide and the dielectrically loaded guide. From the point of view of the transmission line impedance, it is also clear that there is a mismatch between the two transmission lines of different characteristic impedance values.

To develop a transition from a waveguide to fin line, think of the transition as consisting of two different kinds of transmission lines: transition from the empty guide to the partially dielectrically loaded guide and that from the dielectrically loaded guide to the fin line. The dispersion of this cascade of transmission lines can be modeled by a commensurate waveguide uniformly filled with fictitious dielectrics of equivalent permittivities [12–14], as illustrated in Figure 4.3. An application of the perturbation theory yields the equivalent permittivities of each section:

$$\epsilon_{effi} = k_{ei} - (\lambda/2a)^2 \tag{4.14}$$

$$Z_{0i} = Z_{\infty i}/\sqrt{\epsilon_{effi}} \tag{4.15}$$

where $i = 1, 2, 3$. The design parameters for the notch, or the protrusion, is given by ($k_{e1} = 1$):

$$h = b - w = \frac{(\sqrt{k_{e2}} - 1)ab}{\sqrt{k_{e2}}\,(\epsilon_r - 1)s} \tag{4.16}$$

$$k_{e2} = p^2 + [(1 - p^2)(k_{e3} - p^2)]^{1/2} \tag{4.17}$$

where

$$k_{e3} = (1 - (\epsilon_r - 1)\,s/a)^{-2} \tag{4.18}$$

$$p = \lambda/(2a) \tag{4.19}$$

The length of the protrusion, or the depth of the notch, is

$$l = \lambda/4\,(k_{e2} - p^2)^{-1/2} \tag{4.20}$$

where s is the thickness and ϵ_r is the relative dielectric constant of the fin-line substrate.

A basic limitation of the quarter-wave transformer section is that it is narrowband; however, cascade of such transformers can be used to achieve wider bandwidths. This is illustrated in Figure 4.4. The design of maximally flat and Chebyschev transformers is presented elsewhere, and the interested reader is invited to refer to [13].

A simple and practical graphical design procedure for a fin-line taper is presented by Beyer and Wolff [15]. This method requires no knowledge of the electrical parameters of the fin line, rather, it is merely a taper-generation method based on circular arcs, as illustrated in Figure 4.5:

$$r_F + r_H = \frac{1^2 + (b/2 - s)^2}{2(b/2 - s)} \qquad (4.21)$$

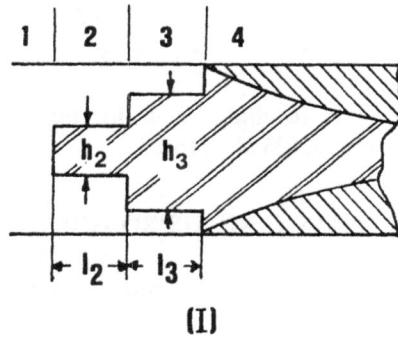

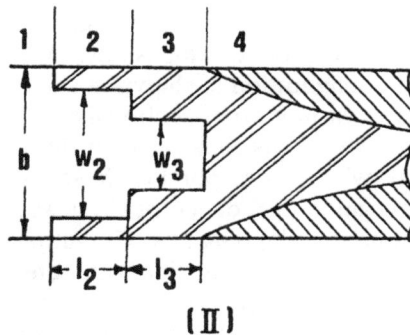

Figure 4.4. Multisectional impedance transformation for matching an empty waveguide to the fin line, using the fin-line substrate as a multistep transformer (after Hoefer and Verner [13], Fig. 1).

62

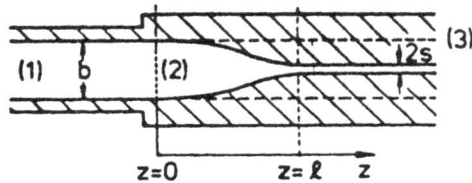

Fig.1: Finline taper.

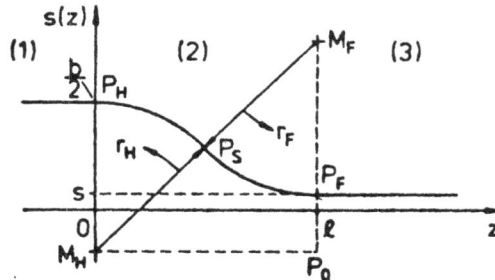

Figure 4.5. A simple way to obtain a fin-line taper by graphical methods (after Beyer and Wolff [15], Fig. 1).

4.4 THE ANTIPODAL FIN-LINE APPROACH

The antipodal fin line is a useful geometry in which a wide range of impedance values can be realized by overlapping the fins at the opposite sides of the fin-line substrate. This is shown in Figure 4.6. The field geometry of this configuration can be rotated 90° to obtain a field geometry similar to that of the microstrip [16]. An impedance match then is achieved by proper design of the fin-line dimensions. A typical taper with the corresponding field configuration is shown in Figure 4.6.

Figure 4.6 shows various cross-sectional cuts of a typical rectangular waveguide to a microstrip. As illustrated in the plane cut views (from A–E in Figure 4.6a), which correspond to those views in Figure 4.6b and 4.6c, an empty waveguide (A), dielectrically loaded waveguide (B), antipodal fin line with nonoverlapping fins (C), antipodal fin line with overlapping fins (D), and antipodal fin line with one fin tapered into the microstrip and the other fin tapered into the ground plane (E). As it is apparent in Figure 4.6c (B), the dielectric slab in the E-plane forces the electric fields to concentrate more in the dielectric. In (C), the metallic fins force the electric field to be perpendicular to the fins. As the fins increasingly overlap, shown in (D), the plane of the electric fields rotates 90°, so that the electric lines resemble those between two parallel plates or a fat microstrip. At this point, a field match between the microstrip and the waveguide has been achieved and an impedance match is needed.

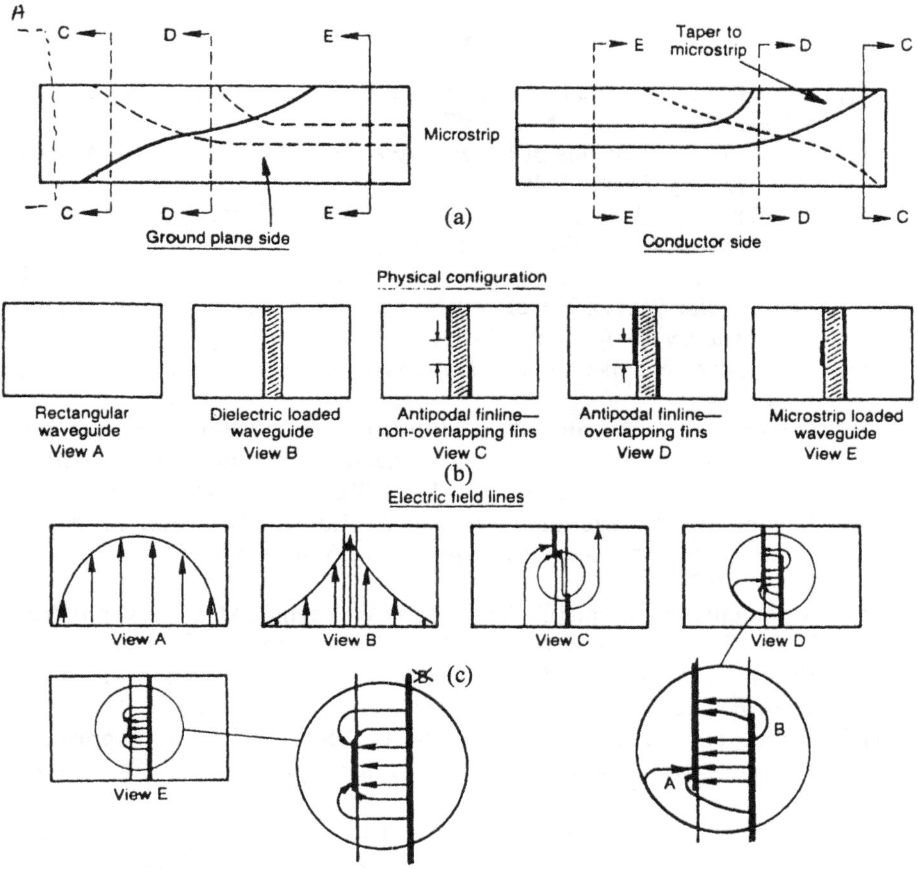

Figure 4.6. Transition from waveguide to microstrip using the antipodal fin-line approach: (a) from the perspective of the transition and various plane cut views; (b) various transmission line media corresponding to plane cut views in (a); and (c) approximate electric field geometries and distributions corresponding to the guides in (b).

The impedance match is achieved by tapering the top and bottom fins to the microstrip and the ground plane, respectively, as shown in Figure 4.6c (E).

The techniques presented thus far can also be used to reduce the reflections at the dielectric substrate interface as it did for the fin line. A more in-depth presentation of fin-line transitions has been presented by Bhat and Koul [17].

64

REFERENCES

[1] Izadian, J., "Unified Design Plans Aid Waveguide Transitions," *Microwaves and RF,* May 1987, pp. 213–222.

[2] Cohn, S., "Properties of Ridge Wave Guide," *IRE Trans.—Microwave Theory Tech.,* 1946.

[3] Hopfer, S., "The Design of Ridged Waveguides," *IRE Trans.—Microwave Theory Tech.,* October 1955.

[4] Pyle, J. R., "The Cutoff Wavelength of the TE_{10} Mode in Ridged Rectangular Waveguide of Any Aspect Ratio," *IEEE Trans. Microwave Theory Tech.,* Vol. MTT-14, No. 4, April 1966.

[5] Marcuvitz, N., *Waveguide Handbook,* McGraw-Hill, New York, 1951.

[6] Hoefer, W. J. R., and M. N. Burton, "Closed-Form Expressions for the Parameters of Finned and Ridged Waveguides," *IEEE Trans. Microwave Theory Tech.,* Vol. MTT-30, No. 12, December 1982.

[7] Utsumi, Y., "Variational Analysis of Ridged Waveguide Modes," *IEEE Trans. Microwave Theory Tech.,* Vol. MTT-33, No. 2, February 1985.

[8] Moochalla, S. S., and C. An, "Ridge Waveguide Used in Microstrip Transition," *Microwaves and RF,* March 1984.

[9] Bharj, S. S., and S. Mak, "Waveguide-to-Microstrip Transitions Uses Evanescent Mode," *Microwaves and RF,* January 1984.

[10] Singh, D. R., and C. R. Seashore, "Straightforward Approach Produces Broadband Transitions," *Microwaves and RF,* September 1984.

[11] Saad, A. M. K., "Analysis of Fin-Line Tapers and Transitions," *IEE Proc.,* Vol. 130, Part H, No. 3, April 1983, pp. 230–235.

[12] Verner, C. J., and J. R. Hoefer, "Quarter Wave Transformers for Matching Transitions Between Waveguides and Fin-Lines," *IEEE MTT-S Int. Microwave Symp. Digest,* 1984, pp. 417–419.

[13] Hoefer, W. J. R., and C. J. Verner, "Optimal Waveguide to E-Plane Circuit Transitions with Binomial and Chebyschev Transformers," *14th European Microwave Conf. Proc.,* 1984, pp. 305–310.

[14] Verner, C. J., and W. J. R. Hoefer, "Quarter-Wave Matching of Waveguide-to-Fin-line Transitions," *IEEE Trans. Microwave Theory Tech.,* Vol. MTT-32, No. 12, December 1984, pp. 1645–1648.

[15] Beyer, A., and I. Wolff, "Fin-line Taper Design Made Easy," *IEEE MTT-S Int. Microwave Symp. Digest,* 1985, pp. 493–496.

[16] Van Heuven, J. H. C., "A New Integrated Waveguide-Microstrip Transition," *IEEE Trans. Microwave Theory Tech.,* Vol. MTT-22, No. 3, March 1976, pp. 144–147.

[17] Bhat, B., and S. K. Koul, *Analysis, Design and Applications of Fin Lines,* Artech House, Norwood, MA, 1987.

Chapter 5
Other Transitions

5.1 INTRODUCTION

This chapter presents some additional transitions, including the new integrated transmission line, called Microslab. In addition, transition to a dielectric image-guide will be discussed.

5.2 SLOTLINE TRANSITIONS

A slotline consists of a narrow gap in a conductive coating on one side of a dielectric substrate, the other side of the substrate is bare. If the substrate's permittivity is sufficiently high, the fields will be confined very close to the slot region and the guide wavelength will be smaller than that of the free space, making it very small in size [1]. The slotline is used in various filter, coupler, mixer, and other semiconductor circuits, and it is rapidly becoming more popular in new applications.

In most applications, a transition is needed between the microstrip and the slotline. Many authors have addressed this problem [1, 2, 3, and 4]. A typical and commonly used configuration is shown in Figure 5.1, where the slotline and the microstrip are positioned orthogonally on opposite sides of the substrate. The crossover is usually a quarter-wavelength from the ends of the respective line. A more complete discussion of the subject and a collection of pertinent references is provided in [4] and so will not be repeated here. The subject is rather straightforward and experimental results using these techniques have been very successful.

A better understanding of the slotline-microstrip coupling can be gained by examining the transition mechanism. Yang and Alexapoulos [5] recently attempted such an analysis, utilizing a method of moments to solve a coupled integral equation. This formulation uses the Green's function for the grounded dielectric substrate, which takes into account all the radiation, surface waves, and surface wave effects. Furthermore, the method of moment solution includes all the mutual coupling effects [5]. This technique could yield a complete characterization of the slot-microstrip line transition over any frequency range.

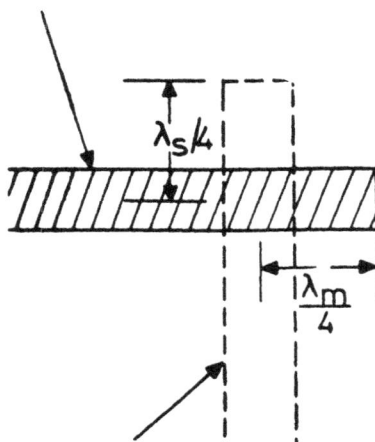

Figure 5.1. A typical microstrip-to-slotline transition (after Gupta *et al.* [4], Fig. 6.5a).

5.3 COPLANAR WAVEGUIDE TRANSITIONS

5.3.1 Coplanar Waveguide to Microstrip

In contrast to the microstrip line, the coplanar waveguide has the advantage of having the ground plane and the signal line on the same side of the substrate. This reduces cross talk between multiple lines in close proximity, a definite advantage. Figure 5.2 illustrates the coplanar waveguide with the advantage of realizing uniplanar circuits not requiring any expensive back-plane circuitry. The gap in the coplanar guide is usually very small and supports the concentrated electric fields. The wider the gap, the larger the impedance will be, and the microstrip will be obtained when the gap is large enough. As the gap reduces, the impedance will decrease, and the impedance will vanish when the gap is zero. Gopinath [6] examines the loss mechanism in the coplanar waveguide.

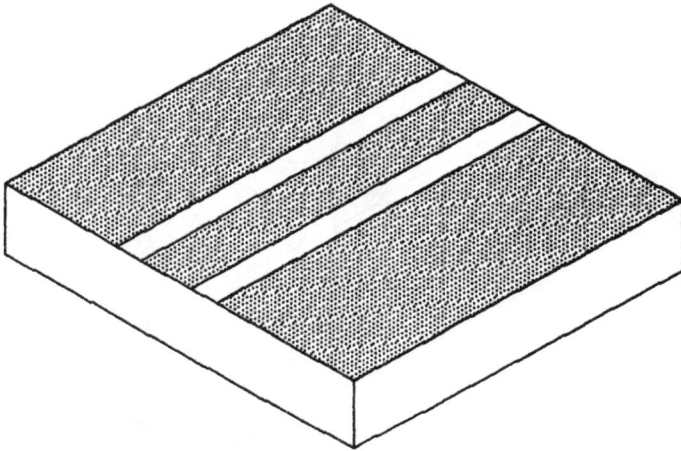

Figure 5.2. A coplanar line and the ground-signal–ground configuration has the advantage of both the ground and the signal on the same side of the substrate (after Gopinath [6], Fig. 1).

In making a transition from a coplanar waveguide to a microstrip, the ground can be brought to the surface of the microstrip through two low-inductance via-holes on either side of the microstrip, as shown in Figure 5.3a. Then, the coplanar line can be started from this point. The via-holes will force the electric field to become parallel to the surface of the substrate. This also is the case for the field in the coplanar waveguide gap. A proper impedance match ensures a well-behaved transition.

Another approach to the coplanar waveguide-to-microstrip transition is illustrated in Figure 5.3b [7] ; here, a two-layer substrate with a ground plan between the layers is used. The microstrip line is realized on the top layer, which is truncated to a single-layer substrate. The coplanar line is then realized on the remaining single layer and a strip connection is used to connect the microstrip to the center strip of the coplanar line, as shown in Figure 5.3b.

These two transitions take advantage of either moving the ground plane to surface and maintaining the same signal line, as with the via-holes in Figure 5.3a, or maintaining the same ground plane and moving the signal line to a lower surface, as in Figure 5.3b. In either case, a connection is needed. A more elaborate scheme is shown in Figure 5.3c, suggested by Houdart and Aury [7]. In this configuration, the signal lines of the two transmission structures are continuous, with small width compensation. The ground plane on the coplanar configuration is realized

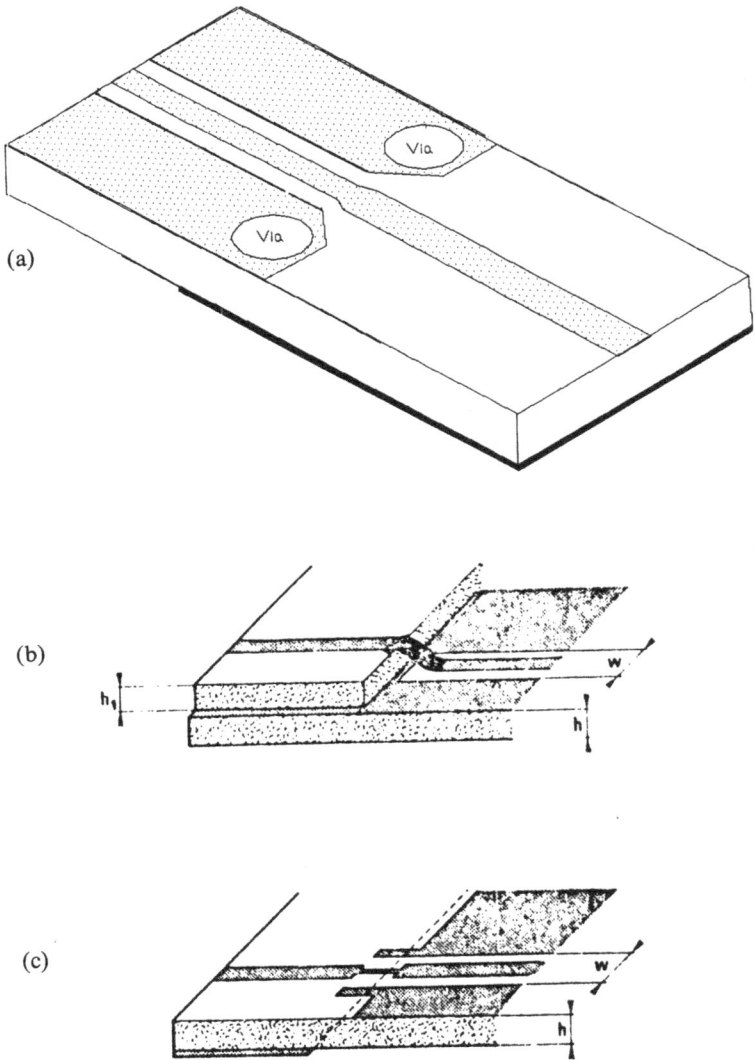

Figure 5.3. Possible transitions from a microstrip to a coplanar waveguide: (a) via holes; (b) the strip connection ground on a two-layer substrate (c) coupled ground transition where no direct ground contact is made (after Houdart and Aury [7], Fig. 2).

through coupling to the ground plane of the microstrip by extending the ground of the coplanar line over the ground of the microstrip, as shown in the Figure 5.3c.

This approach is similar to the realization of a band pass filter constructed by the overlap of the two ground planes [7]. An equivalent model for this approach is shown in Figure 5.4 and suggested by [7]. It is apparent from filter theory that the bandwidth of this transition is dependent on the length of the coupled region, as we might expect. For some useful design insights into an optimum transition of this configuration when a noncontact ground plane transition is desired, we recommend experimental trials and the filter theory.

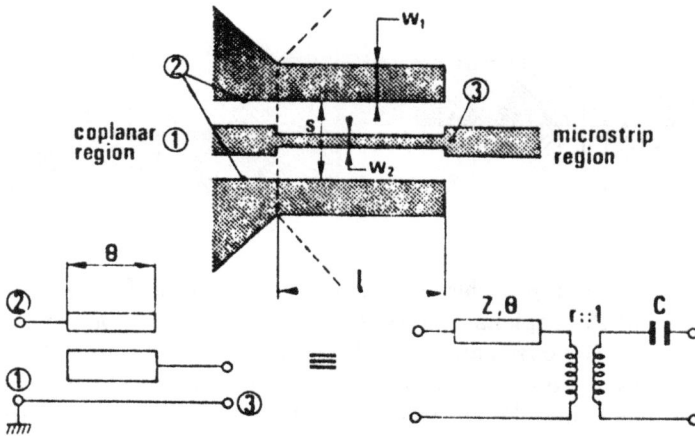

Figure 5.4. Equivalent circuit of the planar microstrip to coplanar line transition.

The transition between coaxial lines and the coplanar line is simple conceptually, and some approaches are illustrated in Figure 5.5. In line transitions are illustrated in Figure 5a, and b. Right angle transitions from coaxial line to coplanar guide can be achieved by using plated-through-via-holes to the back to connect to the coaxial shield and extending the center pin into the substrate under the center contact of the coplanar waveguide (CPW) to make the signal contact. Another approach is just putting the coaxial connector on top of the CPW and making the proper contacts. In each transition just presented, it may be necessary to compensate for impedance discontinuities at the junction.

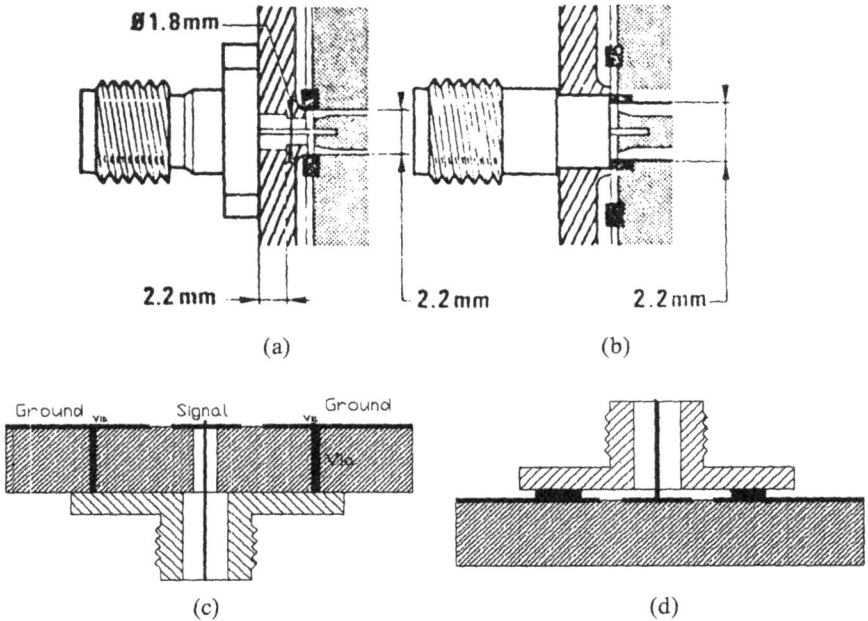

Figure 5.5. Some approaches for making the transition from a coaxial line to the coplanar waveguide: (a and b) transitions as suggested by Houdart and Aury [7] (Fig. 1a); and (c and d) right-angle transitions from coaxial lines to coplanar guides can be achieved by using plated-through-via-holes to the back to connect them to the coaxial shield and extending the center pin into the substrate under the center contact of the CPW to make the signal contact, or (d) by just putting the coaxial connector on top of the CPW and making the proper contacts.

5.3.2 Coplanar Waveguide to Slotline

The transition and interconnection between the coplanar line and the slotline produces some useful configurations. One of these is the classical broadband compensated balun [7], in which the coplanar line is transformed to balanced coplanar strips, as shown in Figure 5.6 with their equivalent circuit model.

An equivalent circuit approach is used by Hanna and Ramoz [8] to analyze the coplanar line-to-slotline transition just described. In this approach, the open-circuited coplanar stub and the short-circuited slotline stub are designed to resonate to transfer maximum power. Figure 5.6b shows the equivalent circuit model for the transition.

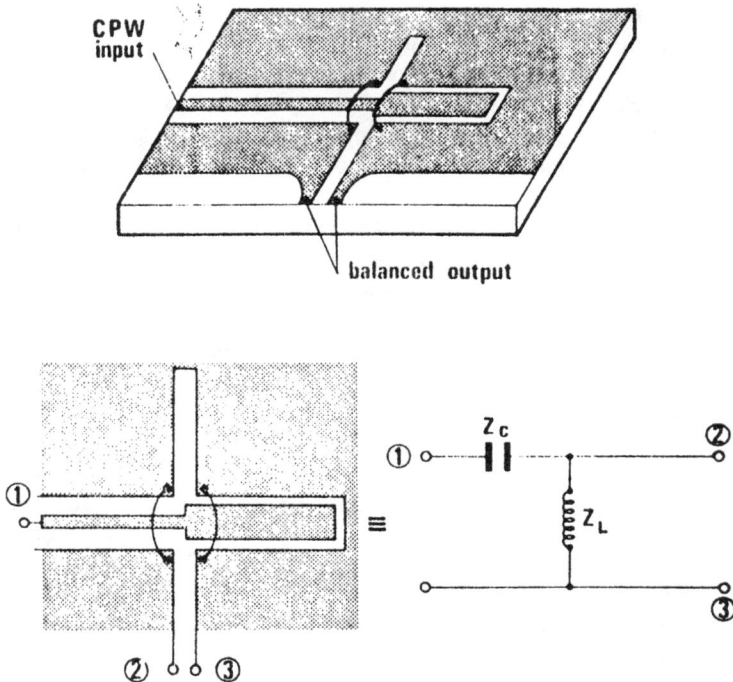

Figure 5.6. A CPW-to-slotline transition that produces a useful balun and its equivalent circuit (after Houdart and Aury [7]).

The optimum length of each stub found from the maximum power transfer is given by [8]:

$$l_c = (\lambda_c/2\pi) \, \tan^{-1}(X_c/Z_{0c}) \qquad (5.1)$$

$$l_s = (\lambda_s/2\pi) \, \tan^{-1}(Z_{0s}/X_s) \qquad (5.2)$$

where X_s is the inductive reactance of a shorted slotline stub of length l_s, X_c is the capacitive reactance of the open coplanar stub of length l_c, and Z_{0s} and Z_{0c} are the respective characteristic impedances of slotline and coplanar waveguide.

A combination of slotline and coplanar line transitions and interconnections using bridging strips, or air bridges, makes a very interesting, useful set of possible circuit configurations for application in MIC and MMIC circuits. Hiroto et al. [9] discuss some useful transitions. The main advantage to these circuit topologies is that they are uniplanar and need no via-holes for ground connections. Ogawa et al. [10] describes an application to the MIC balance doubler. These proposed uniplanar configurations will be discussed briefly because of their potential usefulness and importance. Some useful coplanar waveguide-to-slotline configurations are illustrated in Figure 5.7. These depict transitions from coplanar mode to slotline mode, coplanar mode to coupled slotline mode, and slotline to coupled slotlines.

Useful T-junctions can be realized using the uniplanar configurations shown in Figure 5.7, which can be used in developing hybrid and balanced circuits. Figure 5.7a–c illustrates T-junctions for coplanar waveguide input to either coplanar waveguide or slotlines for in-phase power division. The arrows show the schematic representation of the electric fields in each respective guiding media.

Figure 5.7d–f illustrates the case where the input transmission line is a slotline. Here, each T-junction acts as an out-of-phase divider. The configurations of Figure 5.7g–i show some additional out-of-phase dividers, which have coupled slotline mode inputs, as illustrated by the in-phase direction of the arrows at input port 1. In all the T-junction configurations illustrated in the Figure 5.7, an equivalent transmission line model is provided to help clarify the transmission mechanisms.

A combination of these configurations can be used to realize some very useful hybrid circuits [9]. Three magic-T configurations are proposed in Figure 5.8, where isolation between ports E and H comes from differences between the balanced and unbalanced line characteristics in Figure 5.8a and the noncoupling characteristics of the coupled-slotline orthogonal modes in Figure 5.8b–c.

For insight into the mechanism of these magic-Tees, consider the slotline magic-T in Figure 5.8a, where the fundamental behavior of the magic T can be understood by examining the electric field direction in the slotline and the coplanar waveguide. Figure 5.9 shows a schematic explanation of the circuit behavior. In this figure, arrows represent the schematic expression of the electric field in those lines. Figure 5.9a shows the in-phase dividing performance. The input signal fed to input port H propagates through the coplanar waveguide and is then converted to the slotline by the coplanar waveguide-slotline junction. After propagation through the slotline, the signal is again converted to the coplanar waveguide, and in-phase output signals are obtained from each coplanar waveguide.

The out-of-phase dividing performance is shown in Figure 5.9b. The input signal fed to port E propagates along the slotline. After propagation through the slotline T-junction, the signal is converted to the coplanar waveguide, and out-of-phase output signals are obtained from each coplanar waveguide. The signal does not appear at port H, because the

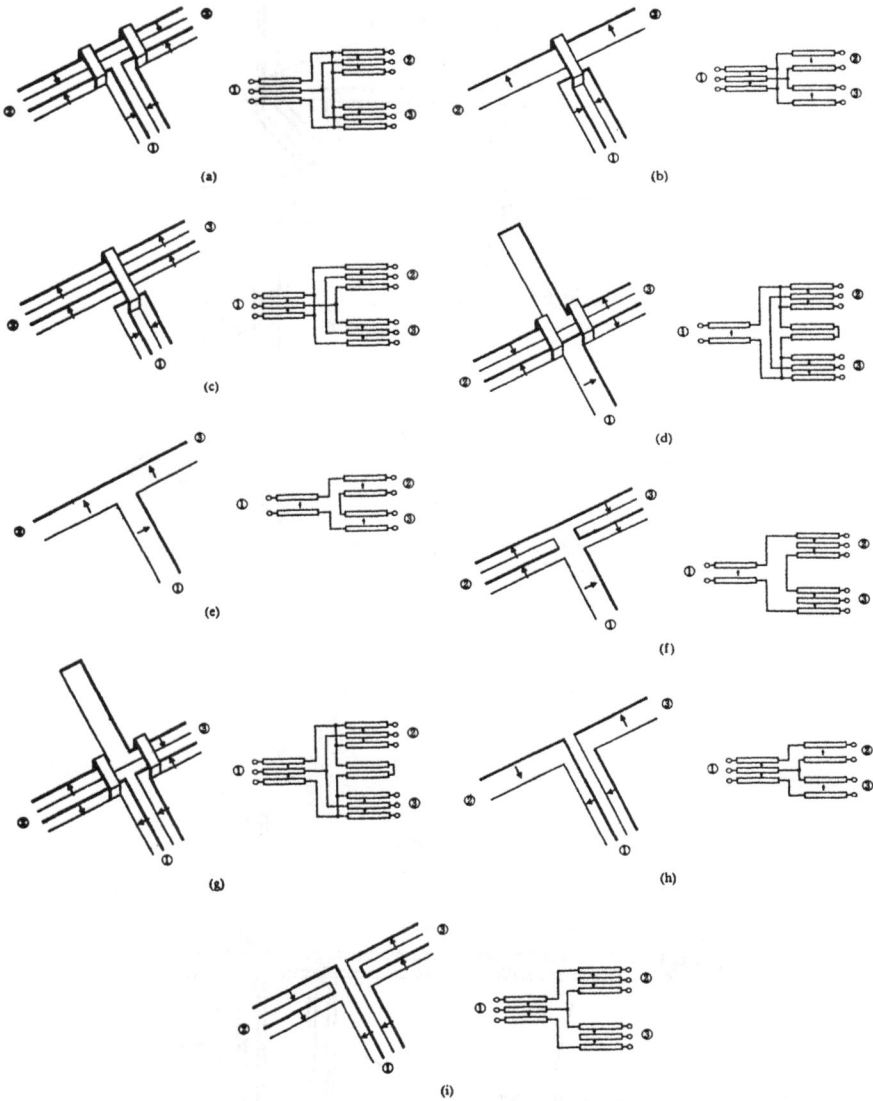

Figure 5.7. Various uniplanar structures useful for MMIC and MIC designs. In each case an equivalent circuit is included to help in understanding the mechanism (after Hiroto et al. [9], Fig. 1): (a) CPW T-junction; (b) CPW-slotline junction; (c) CPW-coupled slotline junction; (d) slotline-CPW junction; (e) slotline T-junction; (f) slotline-coupled slotline junction; (g) coupled slotline-CPW junction; (h) coupled slotline-slotline junction; and (i) coupled slotline T-junction.

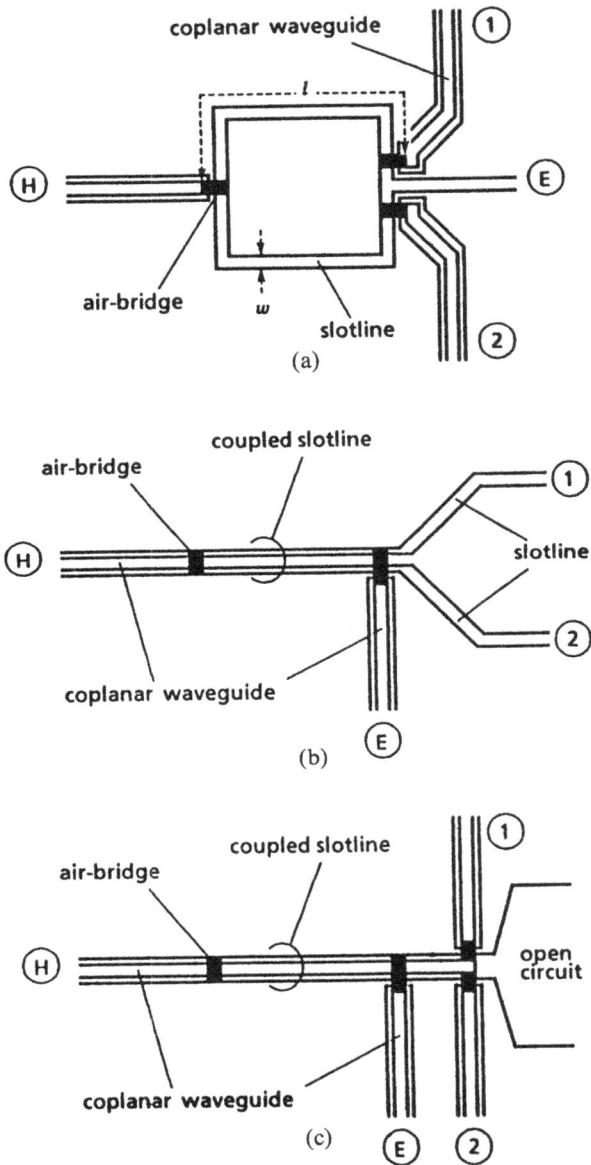

Figure 5.8. Circuit configurations of uniplanar MMIC magic T-junctions: (a) slot line magic T; (b) coupled-slotline magic T with slotline output transmission lines; and (c) coupled-slotline type magic T with coplanar waveguide output transmission line (Hiroto *et al.* [9], Fig. 5).

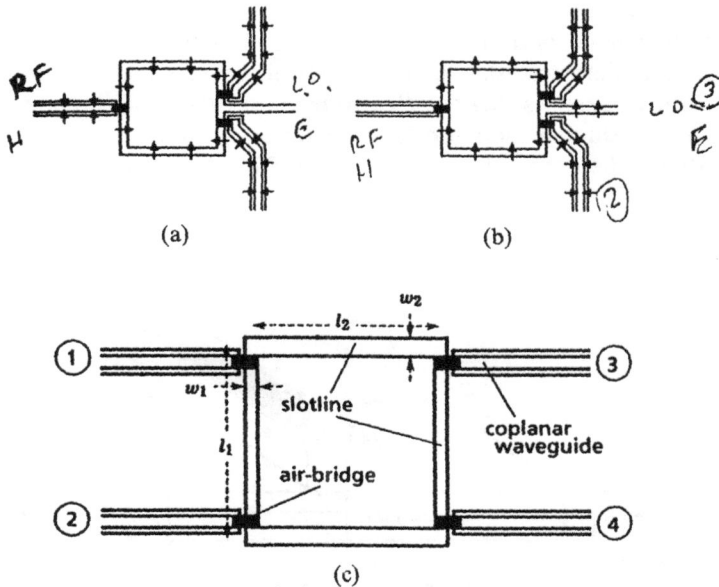

Figure 5.9. Schematic representation of fundamental behavior of the magic T: (a) in-phase excitation; (b) out-of-phase excitation (after Hiroto *et al.* [9], Fig. 6). (c) A branch-line coupler ([9], Fig. 7).

signal that propagates through the slotline is canceled at the coplanar waveguide-slotline T-junctions [9]. A typical realization of a branch-line coupler is shown in Figure 5.9c.

5.4 MICROSLAB WAVEGUIDE

A new semiplanar transmission line, called the microslab waveguide, holds many promises for future applications in MMIC at higher microwave and millimeter frequencies. The term *microslab* is derived from microstrip and dielectric slab waveguide because it possesses the transmission properties of the two media. These properties are achieved by combining the wide band attributes of the metal-bound electromagnetic waves in the microstrip and the low loss of the electromagnetic waves associated with the dielectric slab waveguide. This results in a transmission medium with electromagnetic fields that are tightly bound to the region under the strip, mostly in the dielectric slab, which helps lower the metal strip losses [11].

A simple way of looking at the propagation mechanism in the microslab is to imagine the metallic strip of a microstrip line raised into the air by a dielectric rib support made of, say, alumina. This is illustrated in Figure 5.10. This top metalized rib is then mounted on another dielectric slab of higher permittivity, such as GaAs. Finally, this structure is put on another dielectric slab of lower permittivity, such as alumina. The configuration assures that the electromagnetic wave will be bound on the strip, thus reducing radiation leakage while concentrating it in the high permittivity dielectric slab region, away from the strip, for less Ohmic loss.

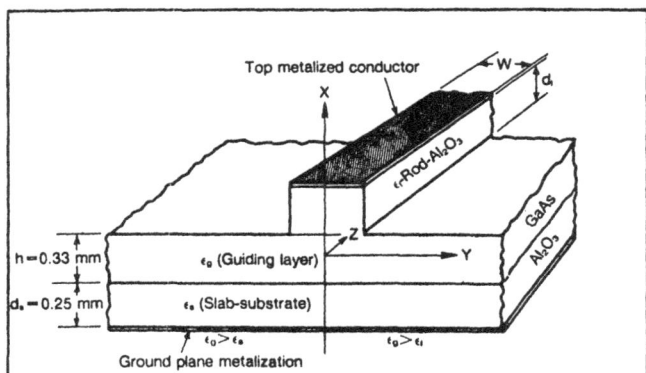

(a)

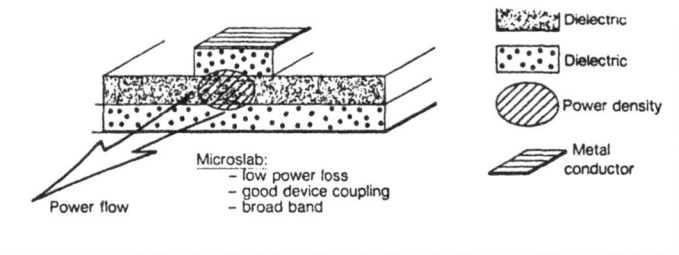

(b)

Figure 5.10. (a) Illustration of the microslab waveguide made from a microstrip by raising the strip and positioning it on a dielectric rib (after Sequeira [11], Fig. 1). (b) The power flow mechanism is shown.

Although the microslab has been suggested for use in the MMIC technologies, perhaps because of its multilayer dielectric, it may be used successfully in MIC technology as the microwave multilayer dielectric for hybrid circuits evolves. Already, attempts have been made to construct a millimeter oscillator circuit [12]. As the microslab becomes more popular in the various circuit designs, efficient transitions to other types of transmission media (e.g., waveguide, microstrip, and slot line) will be necessary. At the time of this writing, however, nothing has been published on this subject.

In making the transition from a microslab and a microstrip line, it is helpful to know that the fundamental field geometries in both are similar, thus only an impedance match is needed. The impedance characteristics of the microslab have been studied [13, 14, 15], which will help in the design of the microslab.

The transition to a waveguide is also fundamentally the same as that illustrated in Chapter 4, for the ridged waveguide-to-microstrip transition. Again, since the field geometries in the microslab and the ridged waveguide are similar, only a proper impedance match is needed.

The transition to a coplanar waveguide can be achieved by truncating the rib and bringing the ground of the microslab to the surface of the middle slab by via-holes then connecting the continuation of the strip into the center strip of the coplanar line, while making grounds on either side of the strip connected by via-holes.

The transition to a slotline is also possible by realizing the slotline on the back of the microslab, as in the transition of microstrip to slotline presented previously.

5.5 RECTANGULAR TO DIELECTRIC WAVEGUIDES TRANSITIONS

A useful transition from a waveguide to a *nonradiating dielectric* (NRD) waveguide is described in [16]. The nonradiating dielectric waveguide is constructed by sandwiching a dielectric strip between two metal plates. This type of waveguide has applications in the millimeter band regions. A typical configuration is shown in Figure 5.11.

Figure 5.11a illustrates the transition from a rectangular waveguide to the NRD. First, a transition is made from the empty waveguide to partially filled dielectric waveguide; then, the partially filled waveguide

78

is transitioned into a dielectrically filled waveguide; and finally, a transition is made from the dielectrically filled guide to the nonradiating dielectric guide by another gradual taper, shown in Figure 5.11b. A cross-section cut at two planes in the partially filled guide and the NRD is shown in Figure 5.11c.

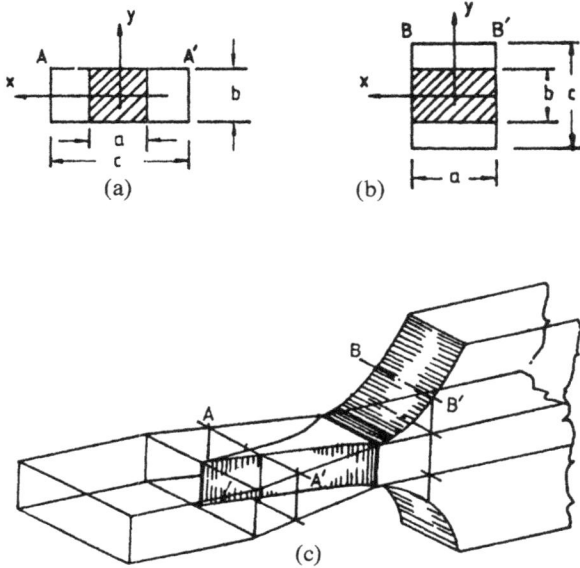

(a)

(b)

(c)

Figure 5.11. (a) nonradiating dielectric waveguide constructed from a dielectric waveguide strip sandwiched between two metal plates; (b) a corresponding transition to a waveguide; (c) cross section cuts for illustrating the transition (after Malherbe *et al.* [16], Figs. 1 and 2).

REFERENCES

[1] Cohn, S. B., "Slotline on a Dielectric Substrate," *IEEE Trans. Microwave Theory Tech.*, Vol. MTT-17, No. 10, October 1969.

[2] Hoffmann, R. K., and J. Siegl, "Microstrip–Slot Coupler Design—Part I: S-Parameters of Uncompensated and Compensated Couplers," *IEEE Trans. Microwave Theory Tech.*, Vol. MTT-30, No. 8, August 1982.

[3] Hoffmann, R. K., and J. Siegl, "Microstrip–Slot Coupler Design—Part II: Practical Design Aspects," *IEEE Trans. Microwave Theory Tech.*, Vol. MTT-30, No. 8, August 1982.

[4] Gupta, K. C., R. Garg, and I. J. Bahl, *Microstrip Lines and Slotlines*, Artech House, Norwood, MA, 1979.

[5] Yang, H. Y., and N. G. Alexapoulo, "A Dynamic Model for Microstrip-Slotline Transition and Related Structures," *IEEE MTT-S Int. Microwave Symp. Digest,* 1987.

[6] Gopinath, A., "Losses in Coplanar Waveguides," *IEEE Trans. Microwave Theory Tech.*, Vol. MTT-30, No. 7, July 1982.

[7] Houdart, M., and C. Aury, "Various Excitation of Coplanar Waveguides," *IEEE MTT-S Int. Microwave Symp. Digest,* 1979.

[8] Hanna, V. F., and L. Ramboz, "Broadband Planar Waveguide-Slot Transition," Centre National d'Etudes des Telecommunications, Division ETR, Issy-les-Moulineaux, France.

[9] Hiroto, T., Y. Tarusawa, and H. Ogawa, "Uniplanar MMIC Hybrids—A Proposed New MMIC Structure," *IEEE Trans. Microwave Theory Tech.*, Vol. MRR-35, No. 6, June 1987.

[10] Ogawa, H., T. Hiroto, and A. Minagawa, "Uni-Planar MIC Balanced Multiplier—A Proposal of New Structure for MICs," *IEEE MMT-S Int. Microwave Symp. Digest,* 1987.

[11] Sequeira, H. B., "Microslab: Waveguide Medium for the Future," *Microwaves and RF,* September 1986.

[12] Sequeira, H. B., and J. A. McClintock, "A mm-Wave Microslab Oscillator," *IEEE MTT-S Int. Microwave Symp. Digest,* 1986.

[13] Sequeira, H. B., J. A. McClintock, B. Young, and T. Itoh, "A Millimeter Wave Microslab Oscillator," *IEEE Trans. Microwaves Theory Tech.*, Vol. MTT-34, No. 12, December 1986.

[14] Young, B., and T. Itoh, "Analysis of Microslab Waveguide," *16th European Microwave Conf. Proc.,* 1986.

[15] Young, B., and T. Itoh, "Analysis and Design of Microslab Waveguide," *IEEE MTT-S Int. Microwave Symp. Digest,* 1987.

[16] Malherbe, J. A. G., J. H. Cloete, and I. E. Losch, "A Transition to Non-Radiating Dielectric Waveguide," *IEEE MTT-S Int. Microwave Symp. Digest,* 1984.

Chapter 6
Microwave Test Fixtures

6.1 INTRODUCTION

In recent years, the development of modern *vector automatic network analyzers* (VANA) has speeded up progress in the field of microwave measurement. Today, automatic, full two-port S-parameter systems regularly analyze networks in up to millimeter frequencies. Several manufacturers are marketing very capable vector analyzers, thus helping to reduce the cost of increased performance.

Taking advantage of the full capabilities of modern automatic network analyzer systems requires a proper test fixture for mounting the device to be tested. The repeatability and the accuracy of the measurement will be a direct function of the performance of the test fixture, making it the most important part of the measurement. For this reason, a considerable amount of effort must be put into evaluating the test fixture, perhaps even to the point of designing a specialized test jig to accommodate accurate and repeatable measurement.

Requirements for test fixtures are as varied as the applications themselves. The diversity of the measurement requirements and the frequency band prohibits designing a truly universal, broadband test fixture. For example, application in the microwave frequencies can be accommodated by a coaxial- or microstrip-based test fixture, whereas measurements in the millimeter regime require waveguide-based fixtures. As a second example of the diversity required, the evaluation of MMIC modules and GaAs digital ICs requires multi-I/O and bias lines, all of which need to be examined carefully and separately, whereas testing transistors alone requires only I/O and two bias lines.

A common denominator among the various applications for the test fixtures is the need for a place to mount the DUT and a design that does not mask or obscure the characteristics of the DUT from the measurement system. Therefore, meaningful and accurate measurements cannot be made without a well behaved, properly calibrated test fixture. In this chapter, we will categorize the various classes of test fixtures and provide a coherent and unified approach in design and development of various test fixtures for microwave and millimeter measurement.

The design of a test fixture is truly a multidisciplinary task. Electrical design is needed to efficiently couple the excitation signal provided by the analyzer to the DUT through proper transitions from one type of transmission line medium to another. Precision mechanical design and innovation are needed to develop an efficient, repeatable, user-friendly mechanical jig. We hope that the materials presented in previous chapters will guide the designer through the electrical and the transition development. The essential innovation and mechanical design know-how can be gained to some degree through experience and feedback from the users.

A test fixture can be represented by functional blocks, as shown in Figure 6.1. These functional blocks include the transition from the VANA ports to the DUT ports. The ports of the VANA are usually coaxial at microwave frequencies and waveguide at millimeter frequencies. The ports of the DUT are usually microstrip for most transistor and MMIC chips and stripline for most packaged transistors and MMIC. The chip-carrier is another functional block within the test fixture, as is the dc biasing network. This simple functional block diagram helps put the design goals in perspective. It is apparent that once the two transmission media are specified, a proper transition can be designed from the information presented in previous chapters. Furthermore, once the DUT is specified as being either packaged or unpackaged (chip), the requirements for the DUT mount are specified and the proper steps can be taken to realize it. The following steps are the logical progression of a typical test-fixture design cycle:

- Specify the requirement of the test fixture: frequency band width, tolerable reflection, insertion loss, isolation, *et cetera.*
- Specify the type of transition needed: coaxial, waveguide to microstrip, *et cetera.*
- Specify DUT type: chip, or packaged, multi-I/O, bias network, *et cetera.*
- Specify the usability requirements: ease of DUT mounting, dismounting, bias connection, accessibility for tuning, *et cetera.*
- Specify methods of thermal management, device cooling, and heat removal.
- Using materials presented in previous chapters, design and implement the necessary transitions.

- Design and implement the mechanical portion of the fixture.
- Test the performance of the fixture by some known standard (such as shorts, opens, throughs, and matched loads) then modify or reiterate the design, if necessary.
- Use the test fixture for actual measurement and evaluated its user-friendliness, then modify and reiterate accordingly.
- Once a fully optimized test fixture is ready, use it in the measurement system and document its design.

Although it is possible to categorize various test fixtures by type, and sometimes by their applications, here we will categorize them by their construction, to help in the clear presentation of the subject.

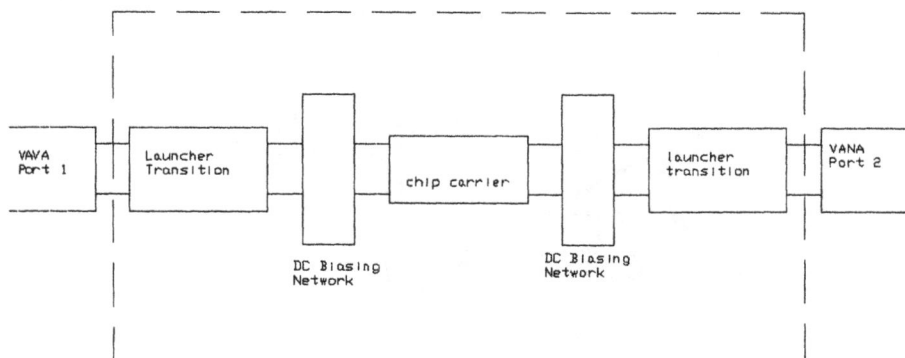

Figure 6.1. Functional blocks of a typical test fixture: the transition-launcher from the measurement system, the chip-carrier, and the biasing and control lines.

6.2 COAXIAL TEST FIXTURES

As mentioned earlier, the output ports of the network analyzers and most other microwave measurement equipment are usually coaxial. This dictates that the test fixtures used with this equipment be coaxial as well. The device under test is either packaged, in which case it will be used in the microstrip or stripline circuit, or in chip form. If the DUT is packaged, a transition from the coaxial line to the stripline is needed. If the DUT is in the unpackaged chip form, a proper chip carrier must be designed to accommodate it. Chip carriers will be discussed in more detail later in this chapter. The transition from coaxial line to microstrip was presented in detail in Chapter 2, and will not be repeated here.

Reference [1] presents a useful and versatile fixture for testing un-packaged chip devices where a transition-launcher is designed to mount and dismount from the body of the fixture, thus making it more modular. This approach is shown in Figure 6.2a. Here, an exploded isometric assembly drawing shows all the various parts of the test-fixture: the two flanged coaxial-to-microstrip launchers, the center block, and the chip carrier including the DUT. Figure 6.2b shows an exploded view of one of the launchers, which is realized by three quarter-wave impedance transformers.

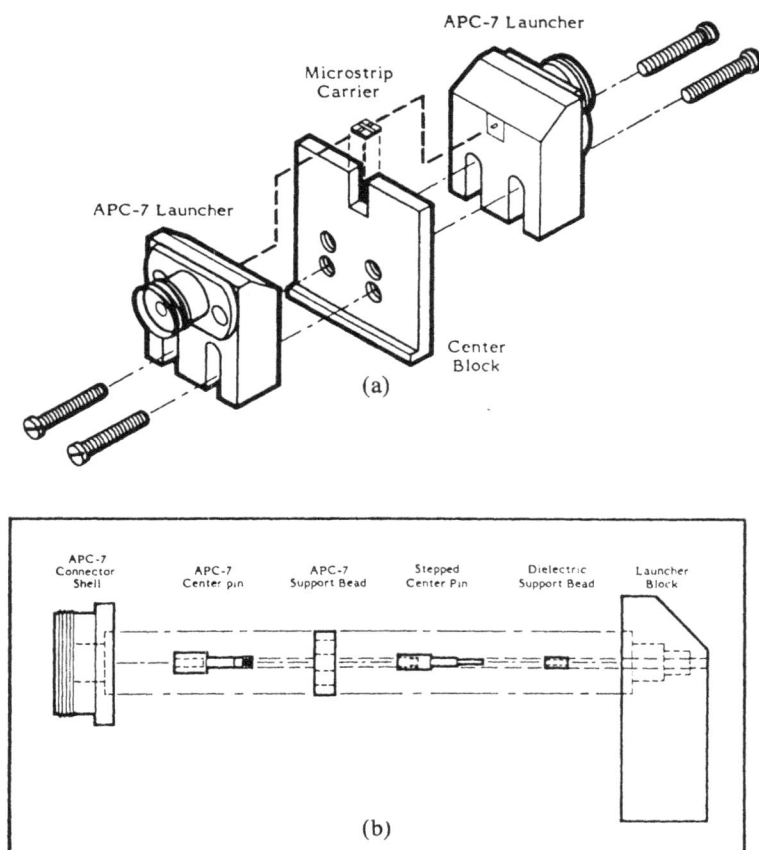

Figure 6.2. A test fixture for testing unpackaged chip transistors and devices: (a) the isometric assembly drawing showing various fixture parts, (b) construction of one of the launcher-transitions (after [1], Figs. 1–4).

Lane and McCollum [2] present a transistor test fixture that extends the capability of this design by partitioning the fixture at the plane of the coaxial line and the transistor interface. As shown in Figure 6.3, the test fixture is composed of various elements: 1L, 2L, 3L in the lower part; and 1U, 2U, and 3U in the upper part. The center part (1L) can be designed to accommodate different package sizes and types or chip carriers for testing chips.

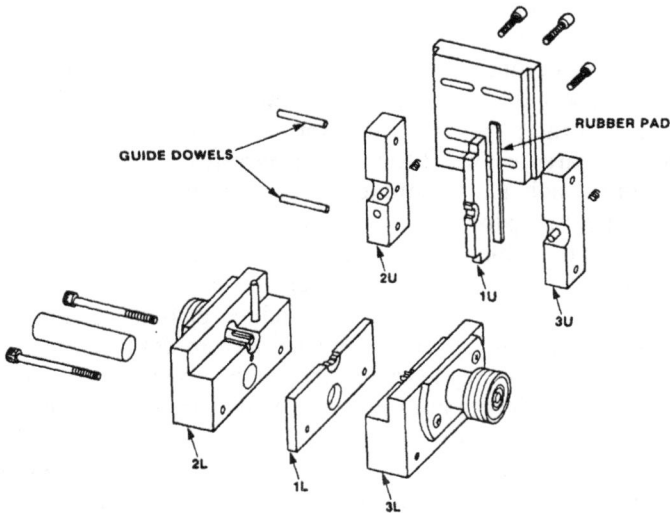

FIGURE 1. Exploded View

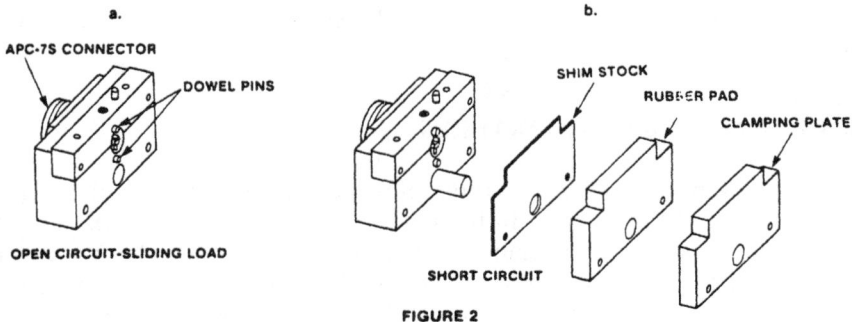

FIGURE 2

Figure 6.3. A test-fixture using fixture calibration standards (after Lane and McCollum [2], Fig. 2).

They suggest a set of calibration standards [2] to bring the reference plane of the test fixture to the packaged device input. The calibration standards include a short, open, through, and sliding load, as is normally used by the automatic network analyzers [3], and can be applied directly at the interface of the coaxial line-transistor interface. A short and open standard are shown in the lower part of Figure 6.3.

A rather simplified variation of the test fixture is presented by Kwaspen [4] and shown in Figure 6.4. The unique feature of this fixture is that it can be manufactured primarily by a turning process rather than milling, thus reducing the manufacturing cost.

As shown in Figure 6.4, a metal plate with a small cut-out containing the transistor insert is sandwiched between two 50-Ω coaxial lines that are connected to the measurement equipment. To test a FET transistor, the source leads are grounded to the two-port insert body, while the gate and drain leads each make contact to one of the inner conductors. A clearance, c, must be provided to prevent an inner conductor-insert short circuit and for repeatable fringing capacitances. The central width, w, of the insert is equal to the gate-drain dimension of the package, and the S-parameters of the packaged device are defined at these planes. The test jig is calibrated with respect to its internal fixture calibration plane. The microwave behavior of the clearance section is modeled by an equivalent lumped element circuit, the parameters of which are calculated from measurement data on a transition-calibration insert [2]. This calibration insert has the same cross section as the transistor insert, but its width is extremely small and its inner conductor has the same geometry as the gate and drain leads of the transistor under test [4].

Other variations of the coaxial test fixture have been reported, each offering a unique feature [5, 6, 7]. They essentially offer the same coaxial line-to-microstrip-stripline transitions. Next we will present the waveguide test fixtures used in the millimeter wave regions.

6.3 WAVEGUIDE TEST FIXTURES

In this section, we present two types of waveguide test fixtures: the ridged waveguide and the fin line. In the millimeter region, the test fixture can be designed using the waveguide-to-microstrip transition. The functional blocks remain the same as was illustrated in Figure 6.1; however, the realization of the transition takes the form of waveguide to microstrip, covered in detail in Chapter 4. The principles discussed there are used to design these transitions. A typical waveguide test-fixture using the ridged waveguide approach of the waveguide-to-microstrip transition is shown in Figure 6.5 [8, 9]. In this approach, the test-fixture is composed of four

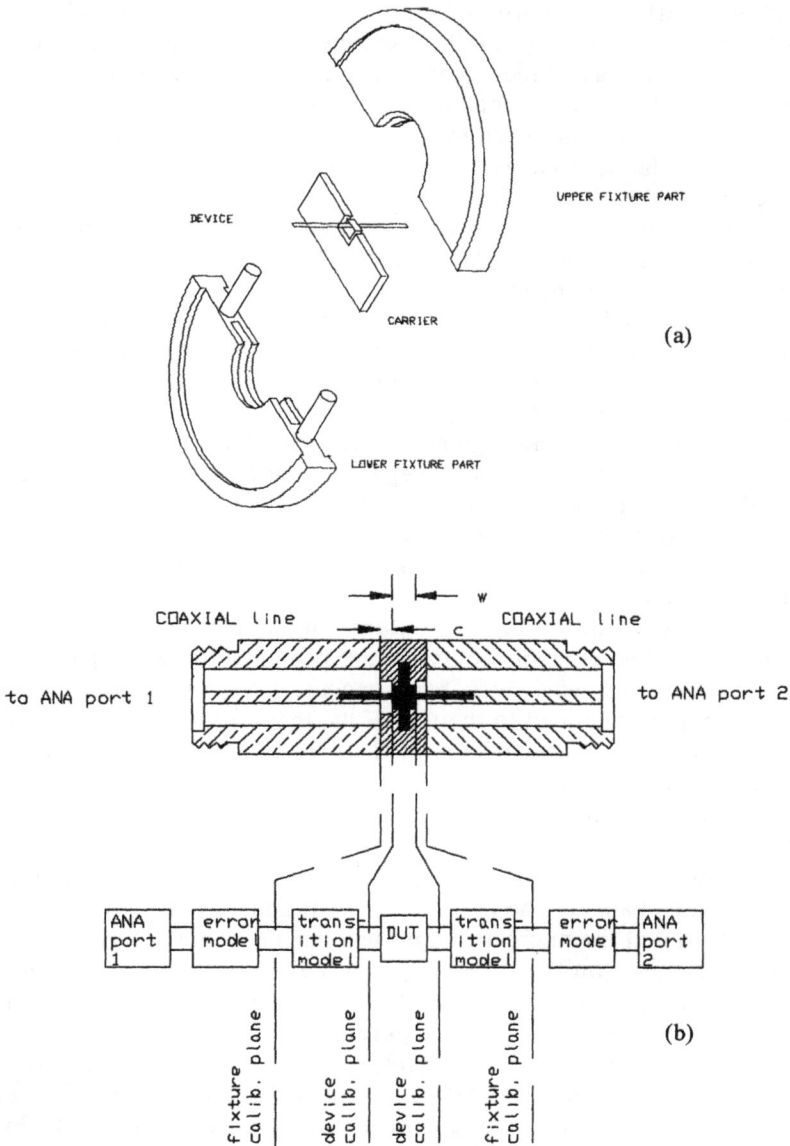

Figure 6.4. A cylindrical construction test fixture similar to that in Figure 6.3: (a) the actual test fixture; (b) a transistor insert (after Kwaspen [4], Figs. 1 and 4).

functional parts, which are designed to be put together and taken apart easily, precisely, and repeatedly. Figure 6.5 shows the four elements of the fixture: two waveguide-to-microstrip launcher-transitions, the center block, the fixture cover, and the chip carrier. In the millimeter region, the transistor is usually used in the chip form, thus eliminating the need for accommodating packaged devices. However, a chip carrier is needed. The detail of the chip carrier will be discussed later.

The launchers or the waveguide-to-microstrip transitions are designed to provide a very smooth transition from the high impedance waveguide to the lower impedance of the microstrip, using the ridged waveguide approach outlined in Chapter 4. The two launchers are designed with special flanges at the microstrip side, to be used for connection to the center block. Proper means for connection to the microstrip chip carrier is also included, in addition to proper alignment slots for ease of use and repeatable connection and alignment. The center block, essentially, houses the chip carrier. It includes the waveguide channel, a small recessed center for holding the chip carrier containing the DUT, as shown in Figure 6.5a, and alignment ridges on either side for fitting the launcher flanges. The fixture cover encloses the chip carrier, functioning as a mode stopper and radiation shield.

Figure 6.5b shows the test fixture fully assembled. The waveguide flanges at the two ends of the fixture are designed to be compatible with the corresponding waveguide band specifications to reduce reflections from the point of contact to the rest of the measurement system. Typical steps to mount the chip carrier in the fixture of Figure 6.5a are as follows. The chip carrier is mounted in the test fixture by removing the fixture cover and loosening the holding screws on either side of the fixture and then easily removing the launchers to expose the channel in the center block. Next, the chip carrier is replaced in the small recessed area in the center of the channel. The chip carrier is designed to fit in this area with precision so that the microstrip lines will line up very accurately with the connecting ridges on the launcher when they are assembled. The fixture cover is then placed in position, the test fixture is then inserted in the measurement system, and the measurement is carried out [8, 9].

Because this test fixture is likely to be used at millimeter frequency bands, special care must be taken to ensure the proper mode-free operation of the test fixture throughout the measurement bandwidth. In the cavity housing the chip carrier, the excitation of higher order modes is possible. The propagation of higher order modes can be suppressed by properly designing the dimensions of the opening in the waveguide channel.

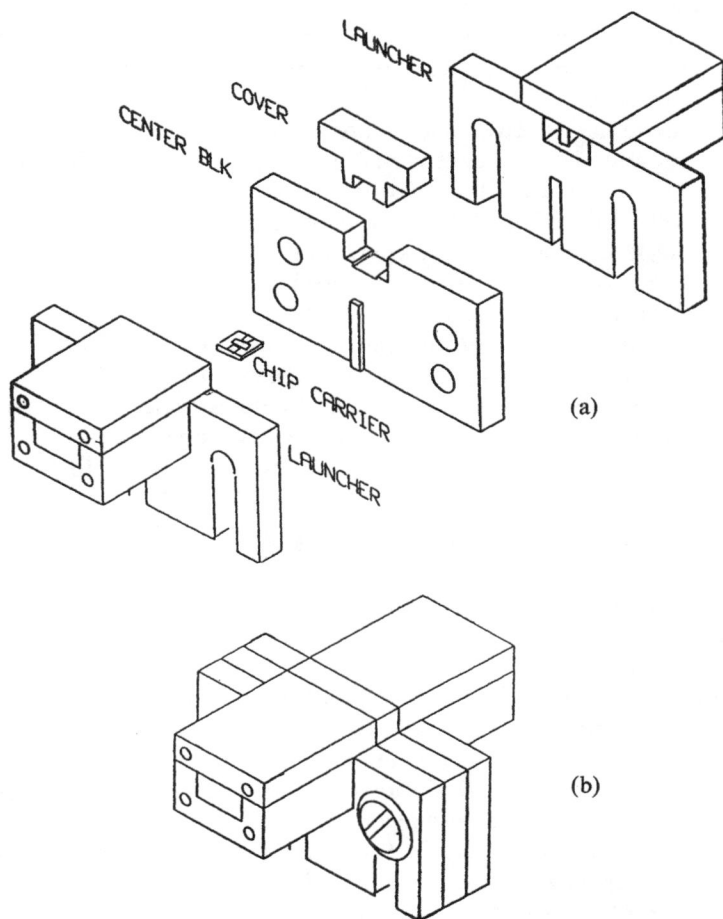

Figure 6.5. A waveguide test fixture: (a) exploded view showing the element of the fixture; and (b) the assembled fixture (after Izadian [8], Figs. 2 and 3).

Another important aspect of this test fixture is the high isolation of the ports in the absence of the chip carrier. This ensures that signal propagation is from the microstrip path and not the direct radiation from one transition end to the other. In order to make sure that no dielectric slab modes are propagating, the chip carrier without a trace is used to check the isolation.

Unique to the ridged transition approach of the test fixture of Figure 6.5 is its integrated biasing network. In this design, the biasing network is realized by dc isolating the ridge from the rest of the fixture body by a very thin covering of insulating dielectric. The dielectric coating covers the entire ridge, except at the tip, where it makes contact to the microstrip. This dielectric coating serves as a by-passing capacitor and dc block, as shown in Figure 6.6a. Its equivalent circuit is shown in Figure 6.6b. In order to bias the device, it is sufficient to connect the dc to the ridges on either side through a choking inductor, as shown in Figure 6.6b. Because the waveguide does not support a center conductor and the ridge also is isolated from the rest of the measurement system, no blocking capacitors are needed, as in a conventional biasing network shown in Figure 6.6c.

6.4 ANTIPODAL FIN-LINE TEST FIXTURES

Another approach to the design of millimeter wave test fixtures is the use of antipodal fin lines. As mentioned in Chapter 4, the antipodal fin line can be used to make a transition from a waveguide to a microstrip line. This makes it possible to develop a test fixture with an integrated transistor mount for measurement of transistors.

Meier [10] suggests using a simple diode mount. A split waveguide block is used to sandwich a thin low-dielectric substrate with a fin line metalization pattern. The split waveguide block is designed to mate with the appropriate waveguide bands, thus the dimensions of the waveguide channel are determined by the corresponding waveguide dimensions. The proper design of the substrate and the taper from the waveguide to the fin line has been covered in Chapter 4. To reduce the reflection at the interface of the dielectric slab and the empty waveguide, a protrusion into the waveguide or proper quarter-wave transformer is designed, as discussed in Chapter 4. After establishing a low-reflection transition between the waveguide and the dielectric slab interface, the metalization is tapered until it reaches the desired gap between the fins for a given impedance value of the fin line. The length of the taper will effect the amount of reflection at the transition from the waveguide to fin line. The taper length

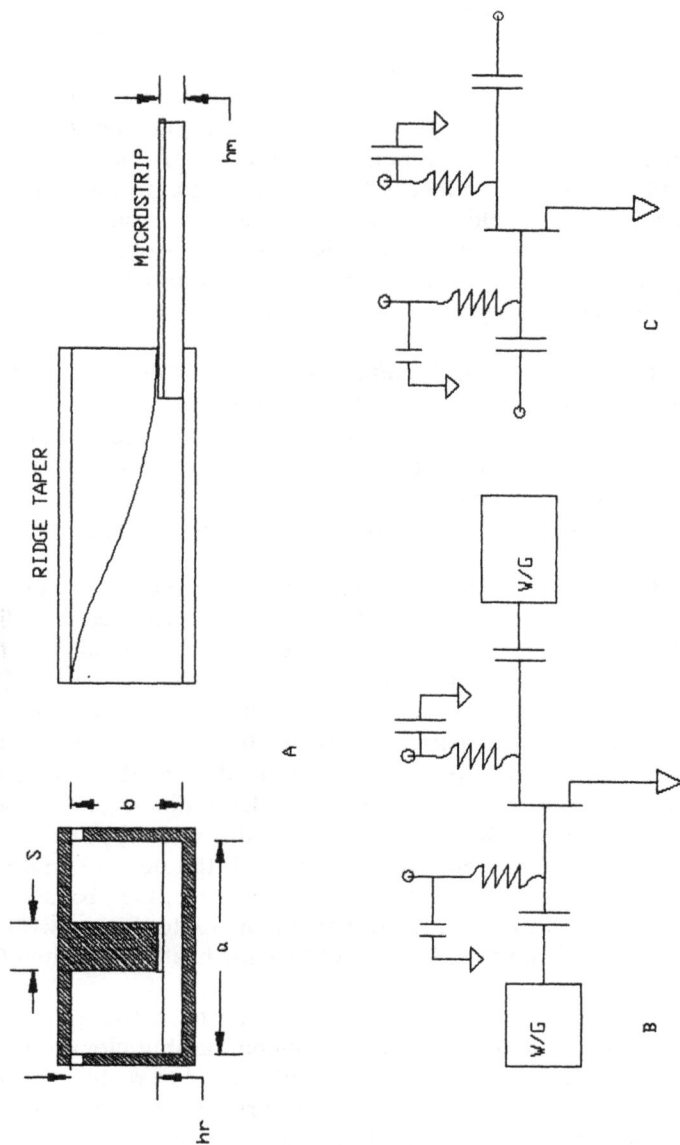

Figure 6.6. Biasing network integrated in the waveguide test fixture of Figure 6.5: (a) realization of the dc block and by-pass capacitor between the ridge and the waveguide structure; (b) equivalent circuit for the bias network; and (c) equivalent circuit of a conventional bias network (after Izadian [8], Figs. 5 and 6).

should be at least a wavelength at the center of the frequency band; keep in mind that the longer the taper, the larger the insertion loss will be. The overall structure can be optimized by accounting for all the trade-offs in terms of size, insertion loss, and reflection.

In the fin-line fixture, dc biasing of the active element can be provided by dc isolation of the fin line's fins from the waveguide split blocks through a very thin dielectric sheet at the point of contact with the waveguide split blocks and an appropriate means for dc connection to the fins [10]. Utilizing nylon screws to hold the split block together avoids short circuiting of the two blocks. Furthermore, it is possible to isolate this structure from the rest of the measurement system by providing a thin dielectric film at the interface with the system port and using additional nylon screws for mating.

The structure of Meier [10] is suitable for two-port devices, such as diodes, but must be modified for use in three-port devices, such as GaAs FETs and other MMICs. For example, to test FETs, the source must be grounded and the gate and the drain are used as input and output, respectively. It is possible to develop a test fixture for the transition from a waveguide to microstrip line using the antipodal fin line approach, as introduced in Chapter 4.

Like the fin-line fixture, a split block waveguide that sandwiches a thin dielectric slab with metalization on the opposite sides is used to realize an antipodal fin line. The interface between the empty waveguide and the dielectric slab is made using a quarter-wave transformer, as before. The transition from the slab-loaded waveguide to the microstrip is made as described in Figure 4.6. The separation between the fins is as wide as the height of the waveguide. This separation is reduced along the taper until the antipodal fins overlap. A squared cosinusoidal, or cubic-type fin taper, can be used [11] to realize this taper. After the fins overlap, one of the fins is tapered into a narrow microstrip trace with the desired width to realize a given microstrip impedance. At the microstrip, the propagation mechanism has changed from a TE_{10} waveguide mode to the quasi-TEM mode in the microstrip and the electric field vector has been rotated 90° (see Figure 4.6).

The biasing network in this approach is similar to that suggested by Meier [10] for the diode mount. Before assembling the split blocks, a thin film of dielectric (mylar or anodized aluminum) is used to realize the test fixture. Nylon screws are used for mating to the rest of the measurement system; a thin dielectric film is added to ensure dc isolation.

6.5 ANTIPODAL FIN LINE VERSUS RIDGED WAVEGUIDE TEST FIXTURES

Although both approaches can provide millimeter wave measurement, some trade-offs are involved in choosing one approach over the other. It is clear that the ridged waveguide taper requires precision, computer-controlled milling to realize the tapers; however, it provides in-line connection to the measurement system (i.e., the transistor will be face up, allowing it to be inspected and tweaked from the top using standard microscopes). In the antipodal approach, the DUT will be rotated 90° and may not be accessible for inspection without modification. Furthermore, the antipodal approach does not lend itself as easily to a fixture with a center block as the ridge type launcher does. Experience has shown that, for the same relative physical size of both approaches, the ridge guide renders lower loss and reflection. Outside of the electrical parameters, the determining factor is the possibility for measurement automation and speed, which can be determined only by the user and the approach.

6.6 CHIP-CARRIER DESIGN

For applications in higher microwave and millimeter frequencies, the transistor is not packaged and is used in discrete chip form to avoid package parasitics. This makes it difficult for testing. However, when the transistor or the MMIC is in the chip, or unpackaged, form, it is possible to provide the means for making I/O contacts to the device. This is done by designing a suitable chip carrier that functions as a platform on which the device is die-attached and connected to microstrip leads via bonding.

There are several approaches to the design of chip carriers. GaAs FET chip carriers need a ground pad, where the source is connected, and two input and output leads for the connection to the gate and drain, respectively. A possible approach is shown in Figure 6.7, where a ground pad is provided by two-plated through-via-holes and two microstrip traces on either side of the ground pad to chip-I/O connections. The transistor chip is die-attached to the ground pad, and the source pads of the transistor are bonded to the ground pad, as shown in Figure 6.7. The gate and the drain of the transistor are bonded to the input and the output microstrip traces, respectively. This approach is very straightforward and can be fabricated in alumina, quartz, or fused silica using standard thin film technology.

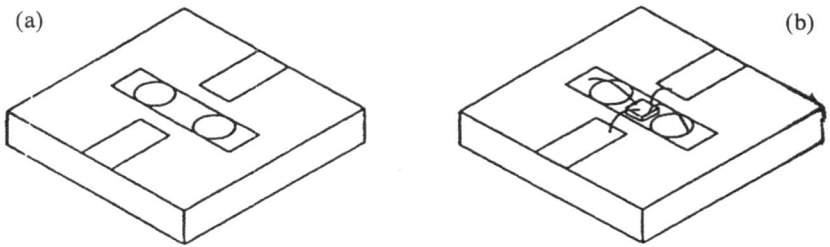

Figure 6.7. A chip carrier using via-holes to bring the ground to the surface: (a) empty chip carrier; (b) transistor mounted on the chip carrier.

A possible drawback to this technique is the requirement of plated through-via-holes for ground pad realization. The inductance of the plated through-via-holes and the bond wires to ground the source (if the sources are isolated from the ground plane of the chip) are likely to form an undesired feedback path, thus degrading the reverse isolation, which could cause oscillation. This can be avoided, however, by optimizing the plated through-via-holes for lowest inductance possible. Furthermore, the thickness of the transistor chip is usually close to the thickness of the microstrip substrate, and longer wire bonds are needed to make contact to the traces and the ground pad, thus causing more parasitics. Additionally, at sufficiently high frequencies, a dielectric waveguide mode may be excited that will bypass the transistor under test and obscure the result. The chosen substrate must be as thin as possible with as low a dielectric constant value as possible.

The ridge-mounted transistor chip carrier, illustrated in Figure 6.8 [11], takes another approach to the design of chip carriers; it uses a machined or stamped metallic chip carrier with a center ridge. The chip carrier is plated with gold, then two microstrip traces are die-attached at either side of the ridge, where the transistor chip is die-attached. The height of the ridge plus the thickness of the mounted transistor chip is usually the same as or slightly less than the thickness of the microstrip substrate used on the sides of the ridge. The gate and drain of the transistor are bonded to the input and the output microstrip traces, as illustrated in the Figure 6.8a. This approach is more expensive, but it exhibits lower and more predictable parasitic effects, as shown in Figure 6.8b, mainly because the plated through-via-holes are eliminated and shorter wire bonds are needed to make gate and drain contacts. This approach also avoids problems with the excitation and propagation of the dielectric waveguide mode, which can be excited in the microstrip substrates on either side because they are isolated by the ridge. In any case, the thinner the substrate, the higher will be the cut-off frequency of these modes.

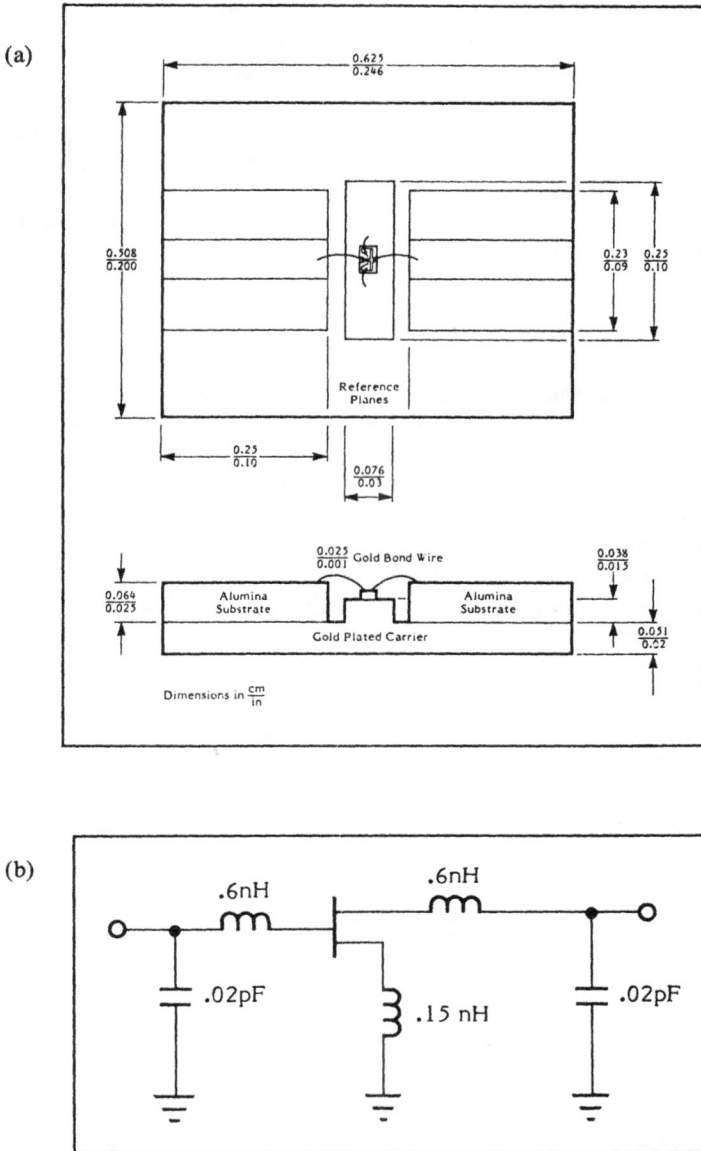

Figure 6.8. A ridged ground pad for the chip carrier: (a) the chip carrier; and (b) its equivalent circuit [1].

Simpson and Lane [12] report a detailed study of transistor chip carriers, which describes the development of chip-carriers to fit a test fixture that is used to test packaged parts. An added feature to this effort was the development of standards that calibrated the test fixture to bring the reference plane to the input of the transistor. This is possible because, unlike the packaged transistor, the chip carrier can be designed to include some calibration standards (such as short, open, loads, and throughs for standard calibration procedures). The technique has been reported to render good results [12].

6.7 DC BIAS NETWORK

Connecting the dc line to the active device is perhaps as important as the RF connection. This is obvious because, without proper dc biasing, the device will not operate at optimum and the measurement data will not be valid. For this reason, special attention must be given to the dc biasing the device and the dc decoupling and blocking in the measurement system. A bad dc biasing network can cause oscillation; and when the device oscillates, no meaningful measurement data can be taken and the problem is very frustrating to solve. A well-designed dc biasing network is the "smart approach."

In coaxial-type measurement, external dc bias networks are readily available and may be provided by the measurement system, such as the *automatic network analyzer* (ANA). Unfortunately, dc biasing networks are not as readily available at higher frequencies and must be designed by the user. In the waveguide test fixtures presented earlier, the bias was incorporated in the test fixture as part of the design. It is also possible to realize the biasing network on the chip carrier. The main element in the chip DC bias network is the DC block. Several researchers have reported on edge-coupled line geometry [13–17].

6.8 TEST FIXTURE EVALUATION AND CALIBRATION

Once a test fixture is developed, it is essential to evaluate its performance to see if it meets the measurement requirements and the design specifications. The test fixture usually is evaluated by measuring through lines or a set of known standards. Under the ideal case, a through line will be the best test of the insertion loss and return loss of the fixture. This through line is made on a chip carrier to simulate conditions similar to those of normal use. In a test fixture for packaged devices, this condition is best simulated by constructing a packaged through line using an empty package or a simulation of the package using the chip-carrier technique.

Perform the through line evaluation in the presence of multiple reflections or ringing in the fixture (i.e., between the two launchers). The noise is likely to appear as some periodicity of nulls and peaks in the measured S-parameter versus frequency. These peaks are inversely proportional to the distance between the two reflection points in the fixture. The further the distance between the reflection points, the closer these ripple peaks will be.

This method is useful when a swept frequency measurement system is available. Additional insights into the location and nature of the reflection points can be obtained by time domain measurement due to impulse or step excitations. High resolution *time domain reflectometry* (TDR) enables us to more precisely locate the discontinuities in the test fixture. Modern vector network analyzers can simulate TDR and, in combination with frequency domain measurement, offer a valuable tool for characterizing and optimizing the test fixtures [3]. A gating feature in the time domain can also be used to isolate some features of the response, such as multiple reflections and higher order mode excitations. Like other useful tools, TDR should also be used with caution and an open mind to avoid misinterpretation and erroneous conclusions.

6.9 TEST FIXTURE CONSIDERATION FOR MMICs

MMIC test fixtures will usually require multiple RF signal I/O and dc control and bias lines. The same technology as described earlier can be used. However, special precautions must be made for providing high isolation between various RF input and output ports as well as dc lines.

The complexity of the test fixture and the special utility of the MMIC fixture dictate a much more careful and well thought-out mechanical design. Benet [18] offers a description of a typical design objective. The fundamental design goal is to develop a fixture system to perform accurate, nondestructive tests on a wide variety of chips, which differ both functionally and physically. The fixture must allow for quick connections and disconnections of the MMIC chip. The fixture also must accommodate a large number of bias inputs without necessitating bonding during the testing process. Multiple RF input-output connections, as well as the multiple bias connections, must also be provided. Provision for monitoring bias voltages is another desirable feature. Metal walls completely enclosing the input-output circuits and the MMIC chip must be provided, to maintain RF shielding and reduce external noise inputs to the device under test. Since a subcarrier would be required, it must be designed for inexpensive production. Finally, since the overall objective is to obtain accurate RF measurement, the fixture must provide a means by which it can be calibrated on an automatic network analyzer [18].

REFERENCES

[1] "Measurement and Modeling of GaAs FET Chips," Application Note, AVANTEK, Inc., October 1983.

[2] Lane, R. Q., and N. McCollum, "A New Self-Calibrating Transistor Test Fixture," *IEEE MTT-S Int. Microwave Symp. Digest,* June 1979.

[3] *HP8510 Automatic Network Analyzer Manual,* Hewlett-Packard Co., Palo Alto, CA.

[4] Kwaspen, J. J. M., "A Coaxial Test-Fixture for Microwave Transistor Characterization," *15th European Microwave Conf. Proc.,* 1985.

[5] DuBois, L., "New Microstrip Fixture and Method for De-Embedding," *MSN and CT,* September 1987.

[6] Ross, P. B., and B. D. Geller, "A Broadband Microwave Test Fixture," *Microwave J.,* May 1987.

[7] Cooke, H. F., "A Universal Fixture for Transistor Chip and Microwave Amplifier Measurement," *MSN and CT,* March 1987.

[8] Izadian, J. S., "Transistor Test-Fixture with Biasing for Millimeter Wave Noise Measurement," *IEEE, MTT-S, ARFTG Conf. Proc.,* June 1987, Las Vegas, NV.

[9] Izadian, J. S., "Measure Noise with a New Fixture," *Microwaves and RF,* October 1987.

[10] Meier, P. J., "Integrated Fin-Line Millimeter Components," *IEEE Trans. Microwave Theory Tech.,* Vol. MTT-22, No. 12, December 1974.

[11] Ebner, H., J. Opfer, and E. C. Schweppe, *A New Integrated Waveguide Transistor Mount,* Institut für Hochfrequenztechnik, RWTH, Aachen, Germany (undated).

[12] Simpson, G. R., and R. Q. Lane, "A Microstrip Chip Carrier and Insert for the Transistor Test Fixture," *25th ARFTG Conf. Digest,* June 1985.

[13] Lacombe, D., and J. Cohen, "Octave-Band Microstrip DC Blocks," *IEEE Trans. Microwave Theory Tech.,* Vol. MTT-20, No. 8, August 1972.

[14] Kajfez, D., and B. S. Vidula, "Design Equations for Symmetric Microstrip DC Blocks," *IEEE Trans. Microwave Theory Tech.,* Vol. MTT-28, No. 9, September 1980.

[15] Tripathi, V. K., Y. K. Chin, and H. Lee, "Interdigital Multiple Coupled Microstrip DC Blocks," *12th European Microwave Conf. Digest,* 1982.
[16] T. Q. Ho, and Y. C. Shih, "Broadband Millimeter-Wave Edge-Coupled Microstrip DC Blocks," *MSN and CT,* April 1987.
[17] Atwater, H. A., "The Design of the Radial Line Stub: A Useful Microstrip Circuit Element," *Microwave J.,* November 1985.
[18] Benet, J. A., "The Design and Calibration of a Universal MMIC Test Fixture," *IEEE MTT-S Int. Microwave Symp. Digest,* 1982.

Chapter 7
De-embedding

7.1 INTRODUCTION

The last chapter presented designs of some test fixtures that could be used to house the DUT for connection and interface with the measurement system (i.e., a VANA or any other equipment). The requirement for the test fixture was to hold the DUT and provide coupling to the I/O ports of the equipment. This usually requires a transition from the ports of the DUT to the ports of the measurement system. Under normal operating conditions, the VANA is calibrated at the I/O ports. When the test fixture containing the DUT is connected to the ports of the VANA, the reference plane for the measurement system is at the ports of the VANA and away from the ports of the DUT. The difference in the reference planes of the device and the measurement system is due to the intervening transition-launcher. It is desirable to move the reference planes to the ports of the DUT.

In an ideal case, where the intervening structure can be thought of as a perfectly matched transmission line, the measurement reference plane could be extended (reflected) to the ports of the DUT by an ideal change in phase of the reflection and transmission coefficients. The previous port extension model serves as a good example for our argument. However, under real measurement conditions, the launcher-transitions rarely can be modeled as ideal transmission lines; at best, this can be used as an approximation or a limiting case, when the launchers exhibit low reflections. Small losses in the launcher-transition can be represented in the transmission line model as a small reduction in the S_{21} amplitude.

In most practical cases, the intervening launcher-transition structures are not so simple and require more elaborate characterization efforts. It is possible to characterize these intervening launchers by systematic measurement of some calibration standards and to deduce the S-parameters for an equivalent circuit that can be used to mathematically move the reference plane to the device port connected to the launchers. This process is called launcher characterization; it has also been called unterminating by Bauer and Penfield [1].

Once the electrical properties (i.e., S-parameters of the intervening structure) are known, the characteristics of the device under test can be de-embedded. De-embedding is the process of deducing the impedance or the S-parameters of a device under test from measurements made at a distance. A theoretical approach to de-embedding, described in [1], is based on unterminating with redundant measurements and iteratively minimizing the experimental errors. Today's analyzers provide sufficient flexibility and data accuracy such that no minimization of the error is required. Thus, most error correction and optimization is carried out in the internal computers of the analyzers.

In this chapter, we review methods of calibration by vector network analyzers. We also present some techniques for embedding and de-embedding the device under test. The chapter ends with a discussion of wafer probing and the future directions in microwave measurement.

7.2 AUTOMATIC NETWORK ANALYZER CALIBRATION

In order to understand the operation and calibration procedure of the automatic network analyzers, it is helpful to refer to the simplified schematic representation in Figure 7.1. As illustrated, the analyzer ports 1 and 2 are where the DUT and the fixture connect. The rest of the system is internal (shown in the dashed box). At port 1, the excitation signal from the swept source incident onto the device is sampled; the reflected signal from the DUT travels back toward the analyzer and is sampled by the coupler. The portion of the signal not reflected is transmitted into the DUT input and altered by the transfer function of the DUT and sent out toward port 2 of the analyzer, and sampled by the coupler at port 2. The reverse of this signal flow is made when the excitation signal is at port 2. In this way, four S-parameters are measured based on the definition given [2]:

$$b_1 = S_{11} a_1 + S_{12} a_2 \tag{7.1}$$

$$b_2 = S_{21} a_1 + S_{22} a_2 \tag{7.2}$$

$$S_{11} = b_1/a_1 \quad , a_2 = 0 \tag{7.3}$$

$$S_{12} = b_1/a_2 \quad , a_1 = 0 \tag{7.4}$$

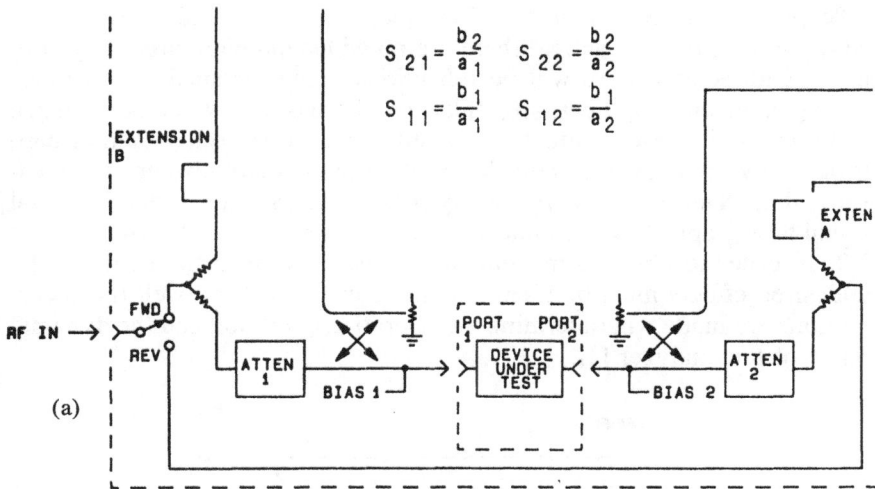

$$S_{21} = \frac{b_2}{a_1} \qquad S_{22} = \frac{b_2}{a_2}$$

$$S_{11} = \frac{b_1}{a_1} \qquad S_{12} = \frac{b_1}{a_2}$$

a. S-Parameter Test Set

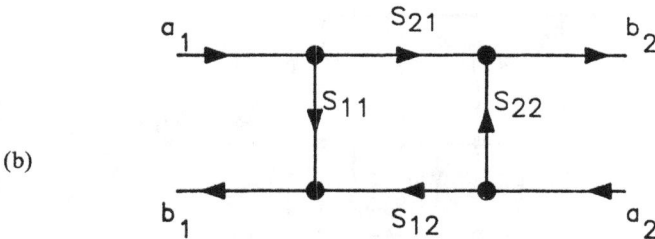

Figure 7.1. (a) A simplified block diagram illustrating the basis of the operation of VANAs; (b) a simple signal flow graph showing the S-parameter relations.

$$S_{21} = b_2/a_1 \quad , a_2 = 0 \tag{7.5}$$

$$S_{22} = b_2/a_2 \quad , a_1 = 0 \tag{7.6}$$

For convenience, a simple signal flow graph is shown in Figure 7.1b. It is apparent that the configuration just described assumes that reflection and transmission are due solely to the DUT and there is no leakage path from port 1 to 2 or vice versa. Furthermore, directivity of couplers is assumed to be infinite. It is further assumed that there are no residual

104

reflections from the connector. This picture is ideal and used only to clarify concepts. In practice, there are no wideband high directivity couplers, and usually there will be reflections, however small, from intervening connectors and perhaps leakage and cross talk, when possible. In addition to all these residual undesired signal flows, other factors contributed by the analyzer and other system parameters obscure the measured data. These effects are usually referred to as error terms. A partial signal flow graph showing some significant error terms is shown in Figure 7.2. In order to obtain a meaningful and accurate measurement, all the undesired effects must be identified and corrected. How well these corrections are made will determine the operating condition and the dynamic range of the analyzer [2].

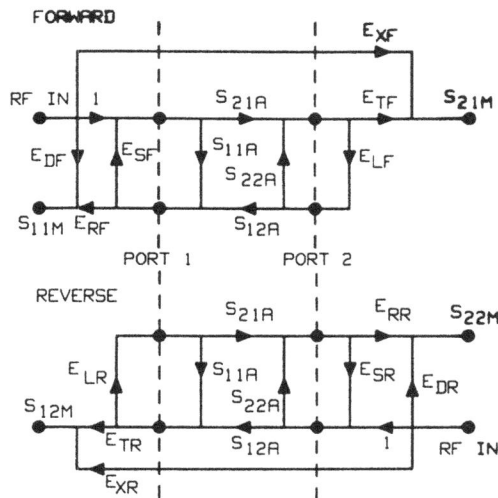

FULL 2-PORT
(S-Parameter Test Sets)

S_{11}, S_{21}, S_{12}, S_{22} Directivity, Source Match, Frequency Response, Load Match, Isolation.

Figure 7.2. Signal flow graph illustrating some possible error signal flows. The S-parameter of the DUT are shown by S, and other error terms are indicated by E[2].

In modern computer-controlled analyzers with internal processing power, it is possible to carry out some systematic measurement of several calibration standards to obtain these undesired parameters and error terms, so that the measured data can be corrected and the true S-parameters de-embedded from the DUT. A detailed presentation of this process, usually called calibration procedure, is beyond the scope of this chapter and the reader is referred to reference [2]. Here, we discuss different calibration procedures as they can be applied.

For a single-port measurement, a fictitious error-two-port can be devised and assumed to be inserted between the measurement port of an ideal, error-free reflectometer and the physical test port at which the unknown reflection is to be measured (see Figure 7.3). This model requires specifications on three independent, complex parameters at each frequency. These parameters are usually called E_{11}, E_{22}, and $E_{12}E_{21}$, which are the elements of the S-parameter matrix of the 2×2 error-two-port. Measuring these three unknown error terms requires at least three equations involving these three unknowns, which can be obtained by measurement of three known calibration standards. A simple model of these single-port measurements with the corresponding two-port-error model is shown in Figure 7.3. It is apparent that a two-port measurement system will provide an additional mirror image of this one-port measurement system and that additional calibration measurements are required to fully characterize the two two-port-error networks. In characterizing the two-port measurement systems some symmetry in the two ports is usually assumed, to considerably reduce the number of unknown error terms and the mathematical complexities. In the following, the OSL, TSD, TRL, and LRL calibration methods will be presented.

One of the most commonly used calibration techniques is measurement of *open, short,* and *load* parameters or OSL for short. The open and short measurements will provide two highly reflective standards with opposite phases. A matched load will provide a reflectionless termination. Sometimes a sliding load is used for better precision. These three measurements will provide three independent equations involving the three unknowns that can be solved to characterize the error networks. In case of the two-port calibration, the measurements are taken for each port, with some additional measurements for port-to-port interaction, like isolation, through leakage, *et cetera.*

A method more suitable for two-port measurement uses *through, short,* and *delay* measurements (TSD), which also provides enough independent equations for the unknowns to characterize all the error terms.

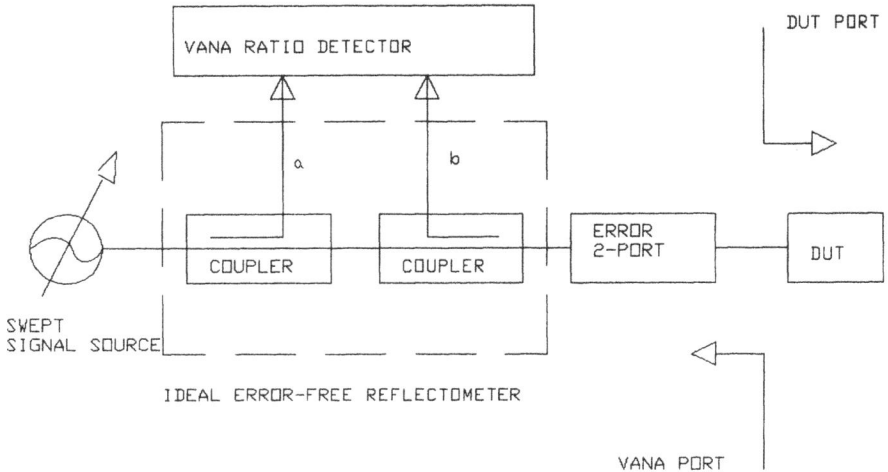

Figure 7.3. A single-port measurement with error model involving an error-free reflectomer cascaded with an error-two-port.

In contrast to OSL, the TSD method requires only two calibration standards in the case of sexless connectors, such as 7-mm coaxial, to specify completely all error terms in the error networks of the full two-port calibration [3].

In the TSD calibration technique, three new two-port-error networks are obtained by the through, short, and delay calibrations. The first calibration measures the cascade of the original two networks in either port; the second is the degenerated short-circuited original two-port-error model, and third is made from a cascade of the original two networks with a section of an unknown length of nonreflecting transmission line sandwiched between them. From the scattering parameters of these three two-port-error-models, closed form expressions for the error-terms are obtained [3].

A method credited to Engen and Hoer [4] utilizes *through, reflection,* and *line* (TRL) characteristics to represent the error-two-ports. Because this method is derived from the TSD, a short-circuit requirement is replaced by an unknown but nonzero reflection standard. This can be a short circuit; for convenience, however, any unknown load could be used. The value of the unknown load can be found as a by-product of the TRL method. The delay line in the TSD remains the same, but the name is changed to *line* to distinguish the method more clearly from the former one. Like the TSD technique, the length of the line (delay in the TSD) can be arbitrary but not close to a half-wavelength. In contrast to the TSD, this line can be lossy [3]. An extension of the TRL method, called LRL, has been proposed by [3]. This method generalizes the TRL by replacing the through requirement with an arbitrary length of a convenient transmission line, thus producing *line-reflect-line* (LRL). The line lengths are arbitrary with their differences ideally being a quarter-wavelength.

A great advantage to this method is that, once calibration is made, the scattering parameters of a two-port network with any combination of connectors can be measured accurately without much change. In contrast, the TSD or TRL technique can be applied only to VANAs having directly mated connectors at the measurement ports [3].

7.3 DE-EMBEDDING AND FIXTURE CHARACTERIZATION

In the last section, we presented some techniques for calibrating VANAs. The calibration procedure involved measuring a set of standards to which the response was well known and, from these measurements, obtaining relations involving the repeatable systematic errors of the analyzer, which had been modeled as two-port-error networks cascaded with ideal error-free analyzer ports. After the two-port-error networks were fully characterized, mathematical manipulation of the error terms, such as inverse cascade chains, is then used to bring the measurement plane to the output of the two-port-error networks. Here, a test fixture containing the DUT is connected. So, at the conclusion of the VANA calibration, the measurement ports will look like the ports of an ideal, relatively error-free analyzer, and any measurement referred to these reference planes are ideal and fully characterized.

In practice, however, the measurement plane and the DUT plane seldom coincide. This is because most DUTs (transistor chips, MMIC chips, MIC circuits) are usually not suited for direct connection to the ports of the network analyzer to be measured; for example, the ports of an analyzer are coaxial whereas most of the circuits have planar strip geometries. These devices are first put in a proper test fixture that acts as a "middle man" or "agent" between the network analyzer and the DUT. This test fixture mates with the network analyzer on the outside with its type of connectors (see Chapter 4 for details of the test fixture design) and connects the analyzer to the DUT in the inside through two launcher-transitions. Therefore, the calibrated network analyzer will measure the scattering matrix of the fixture containing the DUT as a whole and cannot distinguish between the parameters of the DUT and the test fixture containing it.

The scattering parameters of the DUT may be deduced from the overall measured data. This is called de-embedding, as mentioned in the introduction to this chapter. However, before being able to de-embed the device's S-parameters from the measured scattering matrix at the measurement system reference plane, it is necessary to characterize the S-parameters of the test fixture, which is called unterminating [1].

De-embedding is very similar to the calibration procedure of the network analyzer presented in the last section. As you may recall, in the VANA calibration procedure, the two port-error-models were characterized by measuring certain standards. In de-embedding, new networks representing the launcher-transition between the VANA ports and the DUT are assumed and some known standards are measured in place of the DUT. This concept is illustrated in Figure 7.4. Appendix C contains pertinent equations for the de-embedding corresponding to this procedure [5].

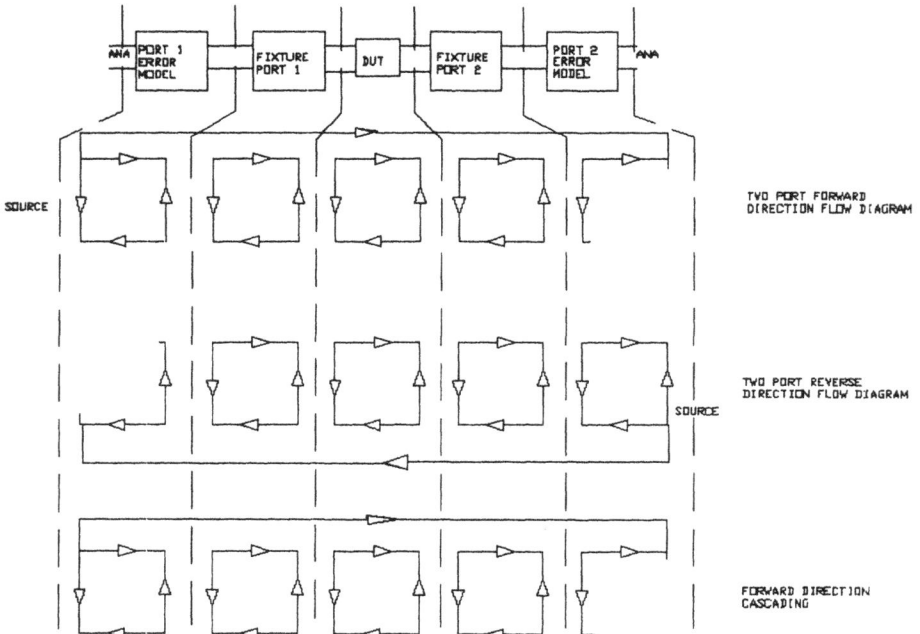

Figure 7.4. The de-embedding procedure and the corresponding signal flow graph for the calibration procedure.

It is possible to categorize two types of de-embedding. First, the VANA is calibrated using its own standards and then fixture de-embedded using the calibration standard in the chip medium. This type of double calibration (one for the VANA, and one for the test fixture) is called a two-tier de-embedding procedure.

It is also possible to perform a single, overall calibration at the DUT level utilizing the proper calibration standards of the DUT medium, this is called one-tier de-embedding. The one-tier method is believed to give more accurate results than the two-tier method, because the propagation of measurement errors is reduced. Some practical consideration such as contact repeatability and contact life, however, favor the two-tier technique [6]. The choices of the de-embedding technique is similar to that of the calibration technique such as OSL, TSD, TRL, and LRL. Some variations of these methods of calibrations can also be used, a good summary of advantages and disadvantages and trade offs of various de-embedding techniques is presented in [6].

Another advantage to one-tier de-embedding is that the analyzer will provide the de-embedded device S-parameter, and no further data processing is required. It is also possible to use, modify, or manipulate the error correction of the VANA to display direct de-embedded data [7, 8]. This ability can also be used to embed a device; that is, insert the device in another, known, network to see the response of the device in the overall system. This technique, which is the reverse of de-embedding, is logically called embedding.

A TSD method has been developed for the microstrip medium by providing precision microstrip standards in a two-tier de-embedding procedure [9].

Staudinger [10] reports that on-chip standards for the MMIC testing enables calibration of the VANA with the on-chip standards in a one-tier TSD method. The technique is suitable for developing standards on the MIC or MMIC circuit for proper de-embedding and calibration. The latter is made possible by wafer probing, which will be discussed at the end of this chapter. A twelve-term error model has been used to develop a de-embedding for a test fixture. The error terms given by [10] are similar to those given in Appendix C. A mathematical treatment of a wideband de-embedding technique is presented by [11]. It uses through, open, and line characteristics, again, similar to the techniques previously presented.

7.4 DE-EMBEDDING IN THE TIME DOMAIN

In past sections, we discussed de-embedding in the frequency domain. That is, the calibration and the measurement of the various calibration standards were made in the frequency domain. The untermination of the launcher-transitions in the test fixture were made by finding the error-model S-parameters as a function of frequency and normalizing the measurement to divide out the measured data by its repeatable measurement errors for various frequencies in the measurement band. Measurement and de-embedding in the frequency domain has many advantages and is very accurate. However, it is limited in the sense that it does not give some direct information, such as the location of sources of various discontinuities. On the other hand, measurements can be made in the time domain, or frequency-domain measured data can be converted into the time domain by an application of inverse Fourier transformation, which can provide more information about the location and the nature of reflection in the measurement setup.

Modern VANAs take advantage of their powerful internal computing power to achieve an inverse *fast Fourier transform* (FFT) of the frequency domain measurement data to display them in time domain; this, in effect, is a simulation of the conventional TDR. In the time domain, various operations (such as multiplication, division, or addition and subtraction) can be made to change for further data analysis and processing. A powerful tool that can be used in the time domain is the gating function, which can isolate various features in time to be Fourier transformed to the frequency domain and compared with the original data. This powerful feature makes it possible to de-embed the launcher-transitions of a test fixture. This de-embedding technique separates the device under test from its environment. The time domain technique is used to directly identify the location of these reflections and allows selection of the window location and its distance. Furthermore, the reflection time domain can be examined and each individual reflection analyzed to indicate the nature of the reflection and the electrical distance to the reflection [12, 13]. Examples of the application of the time domain de-embedding is given in [14] and [15], where a coaxial connector junction between two microstrips is analyzed [14], and a two back-to-back microstrip tapers are analyzed [15]. In both of these cases the utility of the time domain is well illustrated.

Time domain de-embedding, although less accurate than frequency domain de-embedding, provides better insight and clarity. Because it reveals where undesired reflections occur, it gives us an understanding of our measurement inaccuracies. One basic limitation of the time domain is the resolution of two consecutive anomolies in the measurement, which will improve as the frequency of the analyzers increases.

Time domain gating has its greatest accuracy when broadband measurements are made and the gate width is selected to include at least one half-wavelength distance between frequencies. The time domain gate must also begin at the baseline, encompass a complete reflection characteristic, and return to the baseline, if low frequency distortion is to be minimized [12].

The time domain is a tool that can be used in the real-time of measurement to get a quick view of the test fixture in terms of the reflections and losses. Like all other tools, it is useful only when used with care and an understanding of its limitations. We will try to clarify some of the ambiguities about the time domain measurement to aide in obtaining more accurate data and draw correct conclusions.

- The time domain and the frequency domain are related by the Fourier-transform pairs. In numerical calculations, usually FFTs are used and the transformation will always have limitations imposed by numerical transformation pair.
- Response resolution is the ability of the time domain measurement to distinguish between two consecutive, closely spaced reflection points. This is a function of the highest frequency of the measurement system. For equal responses, the response resolution is about the width of the excitation impulse at the -6 dB points. This implies that the response resolution is inversely proportional to the frequency span of the measurement and, therefore, the frequency windowing used (if any) [2].
- Range resolution is a measure of the display equipment used. It is the ability to locate the peak of a single response in time on the display. The range resolution is equal to the digital resolution of the CRT display, which is the time span displayed divided by the number of points. Maximum range resolution is achieved by centering the response on the display and then reducing the time span [2].
- Gating is usually an operation performed in the time domain (time filter). It is the selective removal of reflection or transmission time-domain responses by multiplying the response by a function that is essentially nonzero at the desired response location and zero at everywhere else (i.e., pulse function). This is usually referred to as Gating Function. The effect is that in the frequency domain response of the gated time response due to regions outside the gate is removed. It is helpful to keep in mind that the Gating in time domain is essentially a convolution of the two frequency domain responses of the original time domain response with the transform of the Gating function.

- Windowing is an operation usually carried out in the frequency domain (frequency filter); it is similar to gating in the time domain. (The terms *windowing* and *gating* should not be interchanged.) Windowing helps to reduce ringing in the frequency domain and overshooting in the time domain. Windowing in the frequency domain corresponds to a convolution in time domain.
- *Time domain reflectometry* (TDR) is a technique using a step function to excite the device under test. The reflected response is then monitored by a synchronized sampling oscilloscope. VANAs simulate this operation by using low pass, harmonically related frequency components for which the dc value is estimated. In this mode, the excitation signal can be an impulse or a step function [2]. In contrast to low-pass TDR simulation in VANAs, band-pass relaxes the requirement of having a dc component and harmonically related frequency components. It uses a frequency band and can use impulse or step excitation. Although this is a useful tool, it is not a true TDR.

7.5 DEVICE MODELING

In the past few sections, we presented various methods for de-embedding the device scattering parameters. These various methods help to obtain measurement data at the device reference plans from the data taken at the measurement reference plans. These methods are generally utilized for a packaged device or an unpackaged chip device mounted on a chip carrier, as presented in Chapter 6.

The scattering parameter of the device also may be derived from a totally different technique, namely, by developing an equivalent circuit model based on the device's physical model. The device is fully modeled when all the parameters of this equivalent circuit are determined and, therefore, can be fully characterized for various circuit designs. The term for this is *device modeling*. It is essential, not only because it allows great flexibility in circuit design and conception and in trying new circuit topologies using *computer aided designs* (CAD), but it is also very useful for manufacturing, yield analysis, sorting, and further process development.

Device modeling has been investigated by many microwave semiconductor manufacturers, whose success depends on the development of accurate broadband equivalent circuit models that can be used for all levels of device design, manufacturing, and application. This is because of the complexities and high costs involved in developing new microwave device processes and technologies from research and development to full manufacturing. Computer aided design, *computer aided manufacturing* (CAM), *computer aided testing* (CAT) are truly modern tools to reduce costs and development time and to leverage technical resources. It is clear that none of these computer techniques will be useful unless accurate device models are available for device simulation at various levels, another reason for the importance of device modeling techniques in current and future development of *monolithic microwave integrated circuits* (MMIC).

In this section, we expand on the material developed in earlier chapters and sections to develop a systematic procedure for developing device models. Once a model is developed based on the physical properties of the device, scattering parameters of the device can be calculated based on the intrinsic and extrinsic device parasitics, as illustrated in Figures 7.5 and 7.6. S-parameter measurement of an actual device can be used to verify the accuracy of the model and for suggestions on improving its deficiencies. It is therefore essential to calculate the scattering parameters of the modeled device and correlate them to the measured device at the same reference plane; that is, to either embed the calculated data of the intrinsic device model in the test fixture and compare the measured results or to de-embed the measured parameters at the measurement plane from the measured data of the device in the test fixture to that of the device plane.

Figure 7.5 shows a typical measurement setup where the device is shown as the innermost part of the measurement setup [16]. Other parts of the circuit include the intrinsic device model, the parasitic elements associated with the device, the device package (or the chip carrier parasitic if not packaged), and the test fixture launcher-transitions, which are shown as part of the measurement system. The various levels of a typical GaAs FET have been illustrated and explained, and different ways of obtaining these parasitic is presented in [16].

The transconductance of the intrinsic FET model of Figure 7.6a, gm, along with output resistance, Rd, the gate-to-source capacitance, Cgs, and the gate-to-drain capacitance, Cgd, can be calculated analytically from various FET models presented in the literature [16]. The external parasitic elements associated with the FET chip are shown in Figure 7.6b. The three resistances, Rm, Rdr, and Rf, in series with the gate, drain, and source terminals, respectively, are due in part to the contact resistance of the semiconductor. The capacitors, Cgd_1, Cgs_1, and Cds_1, arise due to the equivalent capacitances between the various metalization layers in the chip.

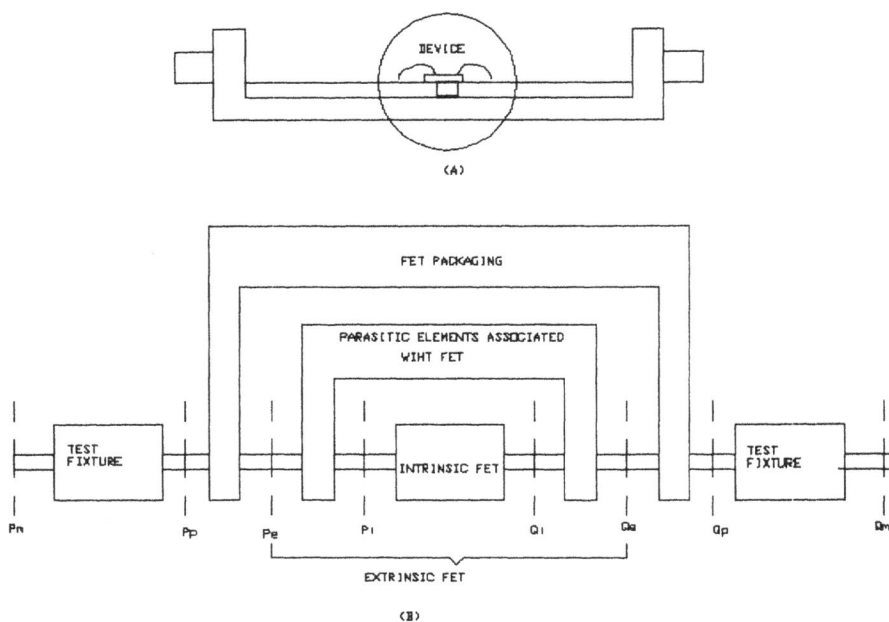

Figure 7.5. (a) Package FET device mounted in a test fixture; (b) portwise models for the intrinsic FET device parasitic, package, and fixture circuit (after Cooper and Gupta [16], Fig. 1).

The package parasitic, or the chip carrier parasitic, is shown in Figure 7.6c. Here, the three sets of parasitic inductors, L_1, L_2, and L_3, represent package lead inductances and C_1, C_2, and C_3 represent the capacitances of various leads to which the chip ports are bonded. Finally, the three inductors, L_4, L_5, and L_6, represent the bonding wire inductances. The various elements of the package can be found by replacing the device in the package by three successive specially designed calibration packages, in which the FET chip is replaced by an open, short, or through circuit between the gate, drain, and source terminals. The scattering parameters of each of the resulting two-ports are measured as a function of frequency, and the package equivalent circuit elements of Figure 7.6c are calculated to fit the measurement [16].

The test fixture model is obtained by the procedure outlines in the previous sections, where a two-tier calibration of a VANA will provide the scattering parameters of the launcher-transitions in the test fixture, as outlined in section 7.3.

115

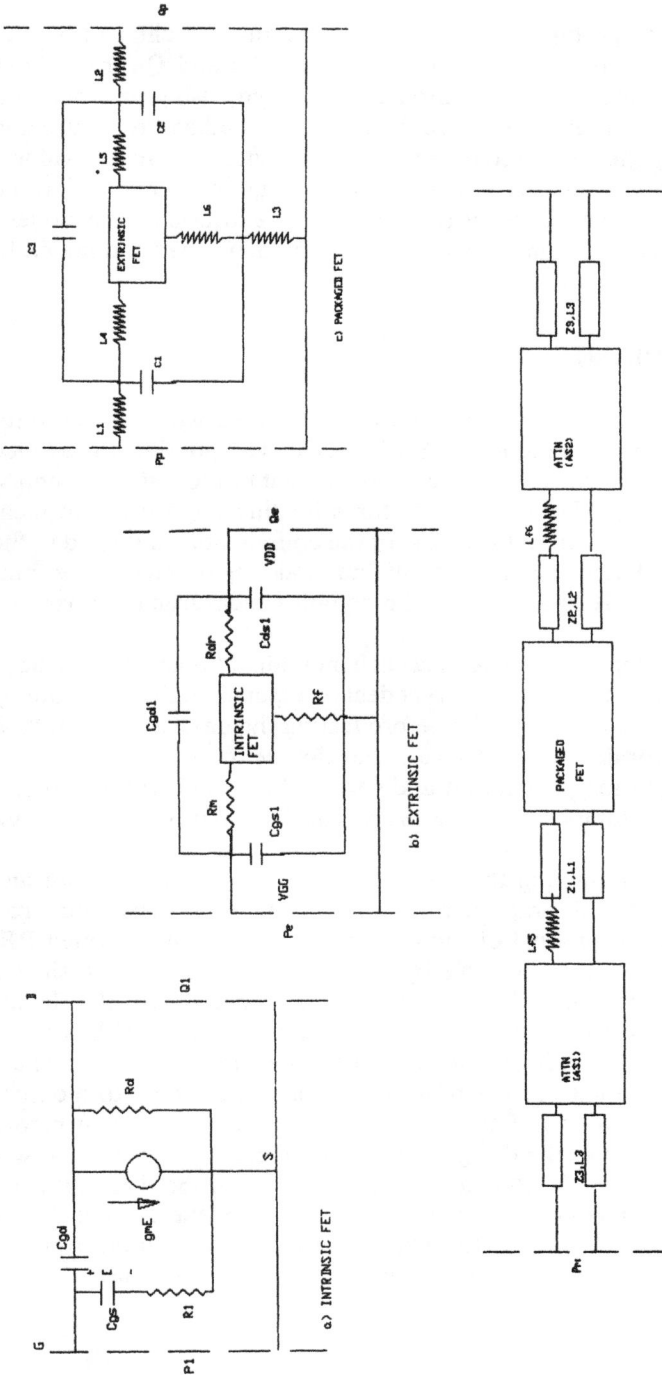

Figure 7.6. Detailed circuit models for (a) the intrinsic FET; (b) the extrinsic FET, which includes the device parasitic; (c) the packaged FET; and (d) the packaged FET mounted in the test fixture (after Cooper and Gupta [17], Fig. 2).

Parameter extraction of the equivalent circuits is achieved by calculating the S-parameters of the device at ports Pe and Qe from those calculated at the measurement ports, Pm and Qm, using any network analysis computer program or cascading and decascading procedure and optimizing-fitting the measured data to various device parameters values. Many different approaches are possible, depending on the end use of the results. All the parasitic values for the intrinsic and extrinsic device models, the package-chip carrier, and the test fixture are determined and the problem is solved [16, 17, 18].

7.6 WAFER PROBING

Combining high speed computer aided de-embedding with the measurement speed and power of modern VANAs has made it possible to advance the state of the art in the microwave measurement of the transistor chips. Development of well designed test fixtures lending themselves to measurement automation will reduce most of the complexities involved in the de-embedding and device measurement and modeling of microwave transistor chips and MMICs and will make complex measurement a routine procedure.

Although complexities in testing each individual transistor have been reduced to a routine task today, a great deal of expensive, time-consuming preparation still needs to be made before testing the device. This method may be suitable for device research and development, but it is inadequate for regular use during production and forces the manufacturers to test only a very small sample and base their yield information on these statistics.

The wafers containing the transistor can be measured before any cutting or packaging. A practice used for dc-probing of these chips can also be used for RF-probing of the die on the wafer. This is called RF-wafer probing. It is made possible by bringing the RF-excitation signals of the VANA to the tips of probes that make contact with the device ports on the wafer. In effect, the RF probes extend the VANA ports to the ports of the probes using the methods described in this chapter. Thus, the de-embedding ability of the probes has made it possible to measure the scattering parameters of the die directly on the wafer even before cutting. This still leaves the design of a proper microwave probe that will couple the ports of the VANA to the device ports on the die. Making a microwave probe thus becomes a logical extension of the current microwave measurement capability, for which a probe is a transition from the coaxial ports of the VANA to the planar geometry of the chip.

Carlton and Strid [19] have described a microwave probe that utilizes two transitions from coaxial line to coplanar waveguide, as described in Chapter 2. The coplanar waveguide is thus tapered to another coplanar waveguide, with the same impedance but much smaller geometry, closer to those of the device level on the wafer. This taper is fabricated on alumina and mounted on a metallic wedge-shaped piece [19]. The transition from coaxial line to coplanar waveguide is achieved by a right-angle transition, as described in Chapter 5. The shape of the coplanar taper and the transition from coaxial to coplanar waveguide is designed and optimized for reduced reflection and loss.

To bring the measurement reference plane to the tips of microwave probes, some on wafer standards can be designed and put on the wafer or special calibrations cards can be made. These calibrations can utilize any of the procedures presented previously, like TSD, TRL, *et cetera*. A one-tier calibration procedure for on-wafer calibration is most logical because good reliable on-wafer standards can be made that are better than the coaxial standards needed for the two-tier calibration [19, 20, 21].

REFERENCES

[1] Bauer, R. F., and P. Penefield, Jr., "De-Embedding and Unterminating," *IEEE Trans. Microwave Theory Tech.*, Vol. MTT-22, No. 3, March 1974.

[2] *HP8510 VANA Manual,* Hewlett-Packard, Palo Alto, CA.

[3] Maury, M., Jr., S. L. March, and G. R. Simpson, "LRL Calibration of Vector Automatic Network Analyzers," *Microwave J.*, May 1987.

[4] Engen, F. G., and C. A. Hoer, "Thru-Reflect-Line: An Improved Technique for Calibration the Dual Six-Port Automatic Network Analyzer," *IEEE Trans. Microwave Theory Tech.*, Vol. MTT-27, December 1979.

[5] Product Note 8510-4, *Equations for Embedding and De-Embedding Device Measurements,* Hewlett-Packard, Palo Alto, CA.

[6] Lane, R. Q., "De-Embedding Device Scattering Parameters," *Microwave J.*, August 1984.

[7] Elmore, G., "De-Embedding Measurements Using the HP-8510 Microwave Network Analyzer," *25th ARFTG Conf. Digest*, June 1985.

[8] Elmore, G., "De-Embedding Device Data with a Network Analyzer," *Microwaves and RF,* November 1985.

[9] Brubaker, D., "Measure S-Parameters with the TSD Technique," *Microwaves and RF,* November 1985.

118

[10] Staudinger, J., "MMIC Tests Improved with Standards on Chip," *Microwaves and RF,* February 1987.

[11] Vaitkus, R. L., "Wide-Band De-Embedding with a Short, and Open, and a Through Line," *Proc. IEEE,* Vol. 74, No. 1, January 1986.

[12] Stinehelger, H. E., Sr., "Discussion of De-Embedding Techniques Using the Time-Domain Analysis," *Proc. IEEE,* Vol. 74, No. 1, January 1986.

[13] Made-It Associates, "Analysis of Transistor Fixture," *Mama's Notes,* Vol. 5, No. 3, May–June 1986.

[14] Made-It Associates, "Analysis of Coaxial Connector Junction Between Two Microstrip Lines," *Mama's Notes,* Vol. 6, No. 5, September–October 1987.

[15] Made-It Associates, "Measuring a Double Taper From 50 to 25 Ohms," *Mama's Notes,* Vol. 6, No. 2, March–April 1987.

[16] Cooper, J. F., and M. S. Gupta, "Microwave Characterization of GaAs MESFET and the Verification of Device Model," *IEEE J. Solid State Circuits,* June 1977.

[17] AVANTEK Application Note, "Measurement and Modeling of GaAs FET Chips," October 1983.

[18] *FetFitter Program Description,* Cascade Microtech, Beaverton, OR.

[19] Carlton, D. E., and E. W. Strid, "Microwave Wafer Probing Achieves On-Wafer Measurements Through 18 GHz," *MSN and CT,* May 1985.

[20] Strid, E. W., "26 GHz Wafer Probing for MMIC Development and Manufacture," *Microwave J.,* August 1986.

[21] Strid, E. W., "mm-Wave Wafer Probes Span 0 to 50 GHz," *Microwave J.,* April 1987.

Chapter 8
Transitions to Antennas

8.1 INTRODUCTION

Antennas can be contrasted to the transmission lines in the sense that antennas are designed to efficiently radiate electromagnetic energy into space and transmission lines are designed to efficiently transport the energy from one point to another without significant loss or dispersion. Antennas can also be thought of as a mechanism of transition from the two forms of the electrical energies, the voltage-currents bound to the transmission lines and the electromagnetic fields freed into the space via the antenna.

The general requirements for transitions between various transmission lines, presented in earlier chapters, can be used in designing transitions to antennas. In this case, however, the transition is from a non-radiating structure to a radiating antenna rather than between two nonradiating structures, as in transmission lines. The transition still must provide an efficient mechanism of field transformation from the transmission line to the antenna structure. Also, the transition itself must not radiate, so as not to disturb the radiation pattern of the antenna with undesired feed-line radiation.

Thus the careful consideration of any transition to be made is an important aspect of the antenna design. In Chapter 1, we outlined the design procedure for typical transitions. A field match was listed as an essential prerequisite for a good impedance match. This is true for antenna transition, too, and much consideration must be given to proper realization of the feed mechanism. A well thought-out transition design and match will make a structure as broadband as possible for efficient coupling of the input power to the radiating structure and will keep to a minimum ringing and wasted stored energies in the near fields of the antenna.

8.2 ANTENNAS AS TRANSITIONS

The simplest and the most common antennas are transitions of well known transmission lines. Figure 8.1 shows the transition from a twin-lead transmission line to a V-antenna and, eventually, to the well known dipole antenna. In the open-ended transmission line (Figure 8.1a), the signal detects an open circuit and part of the signal fringes into the space. The rest of the signal is then reflected back toward the generator. To make an efficient radiating structure, the shape of the wire must be altered to introduce a new geometry, in which the wave traveling in the transmission line does not experience an abrupt change in the impedance and in the boundary conditions or field geometry. This is achieved simply by gradually flaring out the twin leads and smoothly increasing the separation between the current carrying leads, as illustrated in Figure 8.1b. When the antenna structure is realized, it is modified to provide a desired input impedance by varying its length, wire diameters, and flare angle [1, 2].

The V antenna of Figure 8.1b is obtained in the intermediate stage of the antenna's transition from twin-lead transmission line to dipole antenna. The input impedance of this antenna will be higher than the transmission line and lower than the straight dipole of the same length. Numerical methods indicate that, for a 1.5-wavelength V dipole with a flare angle of 118°, the input impedance is about $102 + j14$, as compared to the $125 + j45$ Ω for a straight dipole (flare angle 180°) of the same length and wire diameter [2].

Rigorous solutions are seldom available for practical antennas. Moment method solutions and the thin-wire technique have been used extensively to obtain solutions for many useful structures, such as the V-shaped antennas and others. These powerful techniques can be used to provide input impedance information for matching and insight into the operation and optimization of the antennas. In the absence of these powerful numerical analyses, costly trial-and-error experimentation is essential. Discussion of the numerical methods is beyond the scope of this book, however, the reader is advised to consult Chapter 7 of [2] for an in-depth coverage of the moment method. Another method of considering the input impedance of the V antenna models the V as a finite length of the nonuniform transmission line, with boundary conditions matching on one end, to the transmission line and, on the other, to an open circuit.

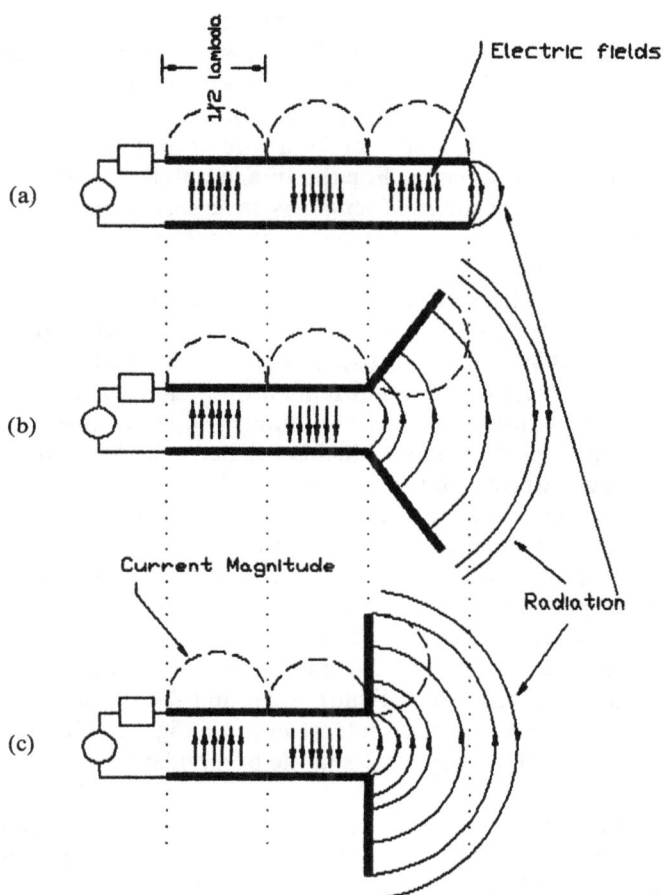

Figure 8.1. Transition and evolution of a dipole antenna from a twin-lead transmission line: (a) open-ended transmission line; (b) transition to a V antenna; and (c) evolution of the dipole antenna (after Balanis [1], Fig. 11).

For a half-wavelength thin-wire dipole, the input impedance is approximately $Z_{in} = 73 + j43\ \Omega$, which exhibits an inductive reactance. Numerical calculations for finite diameter antennas have shown that the antenna length can be shortened slightly to obtain a completely real input impedance; that is, 4 percent shorter for a wire of .001 wavelength radius to provide a resonant antenna. For the dipole antennas with finite wire diameters, the resonant antenna length is inversely proportional to the wire diameter. This implies that thinner wires will require less trimming of the antenna length to obtain resonance [2]. The dipole antenna was shown earlier to be a transition from a twin-lead transmission line, which suggests that the best way to feed it is with a twin lead. In practice, however, a coaxial line usually is used. The incompatibility of the current flows in the two structures thus presents a problem. The coaxial line to dipole transition is unbalanced, and without proper compensation, unbalanced currents will flow on the shield of the coax and the antenna arm. The unbalanced currents flowing on the dipole antenna distorts the radiation pattern and causes impedance mismatch problems that must be eliminated. This is commonly handled with a balun.

Figure 8.2 illustrates the horn antenna, another popular antenna that can be used as an example of transition from a transmission line. In this illustration, the open-ended waveguide already acts as a radiating structure at the open end. Flaring out the sides of the guide H-plane will result in the H-plane sectoral horn of Figure 8.2a, and flaring out the sides of the E-plane makes the E-plane sectoral horn of Figure 8.2b. The simultaneous flaring of both planes results in the pyramidal horn of Figure 8.2c [2].

Assuming a dominant TE_{10} operation, the impedance seen on the waveguide side is essentially that of the regular waveguide, provided that the transition to the horn is relatively reflectionless and does not excite higher order modes. In this case, the waveguide impedance definitions given in Chapter 1 can be used to define proper matching impedance for any of transitions shown in Figure 8.3.

Figure 8.3 illustrates some typical horn antenna feeds. Figure 8.3a uses a coaxial probe transition, presented in Chapter 3 and discussed in detail; Figure 8.3c uses a magnetic probe or loop antenna, presented in Chapter 3; also Figure 8.3b illustrates the feed using a microstrip, and transition to the waveguide by a single-ridged waveguide, presented in Chapter 4; Figure 8.3d illustrates the horn feed as realized by the transition from microstrip to antipodal fin line or fin line to waveguide, discussed in Chapter 4. Finally, Figure 8.3e illustrates a microstrip probe transition that is treated similar to the coaxial probe of Chapter 3.

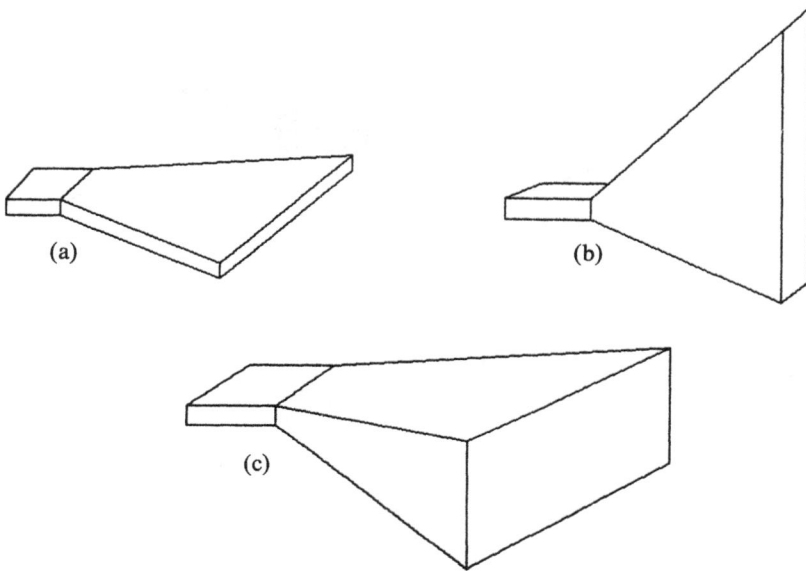

Figure 8.2. Horn antenna as a transition from an open-ended waveguide: (a) flaring out the H-plane produces the H-plane sectoral horn; (b) flaring out the E-plane produces the E-plane sectoral horn; and (c) flaring out in both planes simultaneously produces the common pyramidal horn.

Determining the input impedance for the horn antenna in a more complex case, as when reflections and higher order modes are present, is rather tedious and possible only with accurate determination of the total internal field in the waveguide at the impedance point. This complex field will be a superposition of the reflected fields, difracted fields, and the higher order fields at the impedance point. It may be possible to determine the points of reflection and apply geometrical ray tracing to fully characterize these fields. The diffracted rays are likely to be due to the abrupt discontinuity of the rim of the horn antenna, and *geometrical theory of diffraction* (GTD) can be used to identify and trace these rays to the impedance point. Once the total fields are determined, the impedance can be obtained by any of the three impedance definitions given in Chapter 1: power-voltage, power-current, and voltage current.

124

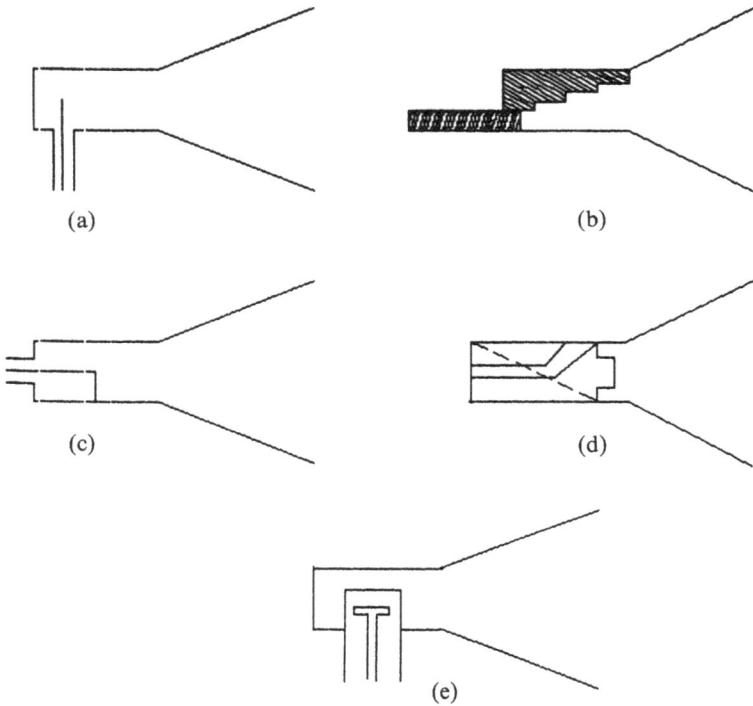

Figure 8.3. Typical horn antenna feeds: (a) using a coaxial probe transition; (b) ridge transition from microstrip; (c) loop feeding; (d) anti-podal transition from microstrip; and (e) microstrip probe feed.

8.3 MONOPOLE ANTENNAS

A monopole antenna is easily realized by extending the center conductor of a coaxial line while spreading the coaxial line shield into a ground plane, as illustrated in Figure 8.4. This antenna will have a balanced current flow and will make an efficient radiating pattern in horizon. In an ideal case, the ground plane is assumed to be infinite in extent; however, in practice, the ground plane is finite in dimensions. This alters the performance of the antenna and its radiation pattern because of current reflection at the edge of the finite plate or diffraction or both. This effect can be modeled and accounted for by geometrical diffraction theories [2], the method of moment [3], or a hybrid of these two techniques.

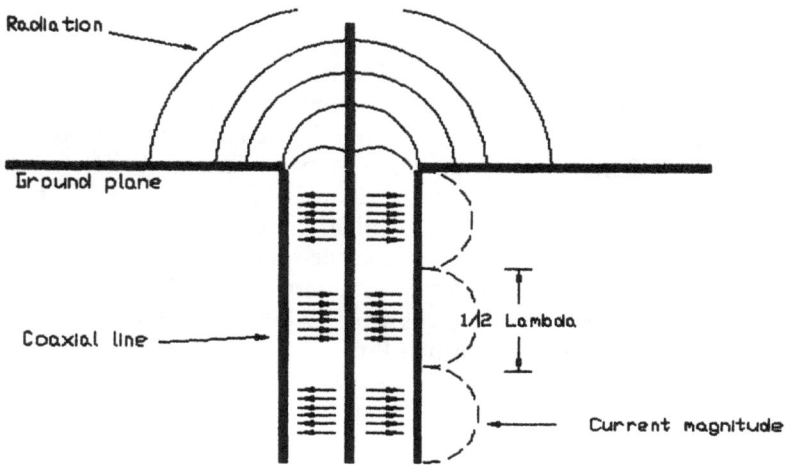

Figure 8.4. A monopole antenna can be made from a coaxial line by extending the center conductor and spreading the shield into a ground plane.

For thin-wire antennas, the input impedance of a monopole antenna on an infinite ground plane will be half that of an equivalent dipole antenna of twice its length. This can be seen from an application of the image theory. In practice, the ground plane will have finite dimensions and the wire will have a finite thickness, thus, the input impedance of the monopole antenna will be a function of the size of the finite ground plane. This is mainly because the currents at the center of the circular disk on which the monopole is located form the superposition of the original coaxial currents and the reflected currents arriving from the plate edge.

In a moment method approach, Richmond [3] utilizes a sinusoidal-Galerkin technique for modeling a monopole centered on a circular disk. This technique is used to provide a set of plots for the input impedance of a monopole as function of the disk radius. These calculations suggest that a thin monopole antenna with a wire diameter of .006 wavelength and a length of .229 wavelength will have a resonant input impedance of approximately 31 Ω for a circular disk of radius of about .45 wavelength. In this particular case, the antenna exhibits a capacitive reactance for larger size disks. This impedance, however, will be affected by varying the monopole height, wire radius, and the disk size. It is likely that the input reactance will become inductive close to a wire length of a quarter-wavelength for this case.

126

8.4 PRINTED ANTENNAS

Printed circuit dipoles can be realized on microwave substrate. An approach to feeding a bow-tie printed dipole antenna is given in [4]. Here, a transition from microstrip to slot line is used advantageously to realize a balun from the unbalanced microstrip line to the balanced coplanar strips, which are realized by transitioning slot lines to coplanar strips as shown in Figure 8.5 with its equivalent circuit. In making the transition from the microstrip to slotline, quarter-wave transformers are used to provide for an impedance transformation of 4:1. The values for these transforming section impedances are 57.5 and 34 Ω on the microstrip and 154.3 and 89.5 Ω on the slot line [5].

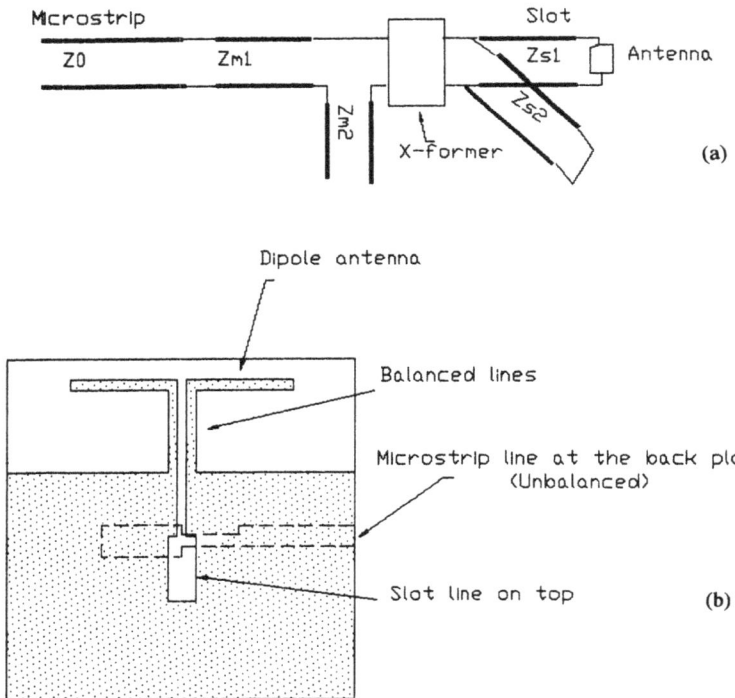

Figure 8.5. A balun realized by a transition from microstrip to slotline: (a) the equivalent circuit; (b) the application to a printed dipole feeding (after Nesic [4], Figs. 2 and 4).

Strip dipoles can be made from a transition of a coplanar strip, as shown in Figure 8.6a, and a slot dipole from the transition of a coplanar waveguide, as illustrated in Figure 8.6d. These antennas can easily be made into bow-ties for more broad banding, as illustrated in Figures 8.6b and 8.6e. The strip and the slot dipoles are complementary and their input impedances are related by a simple expression [6]:

$$Z_s = \eta_0^2/(4Z_d) \tag{8.1}$$

where Z_s is the slot impedance, which will be of an order of several tens of ohms, η_0 is the free space wave impedance, and Z_d is the antiresonant strip dipole impedance of an order of several hundreds of ohms, which can be determined by numerical calculation.

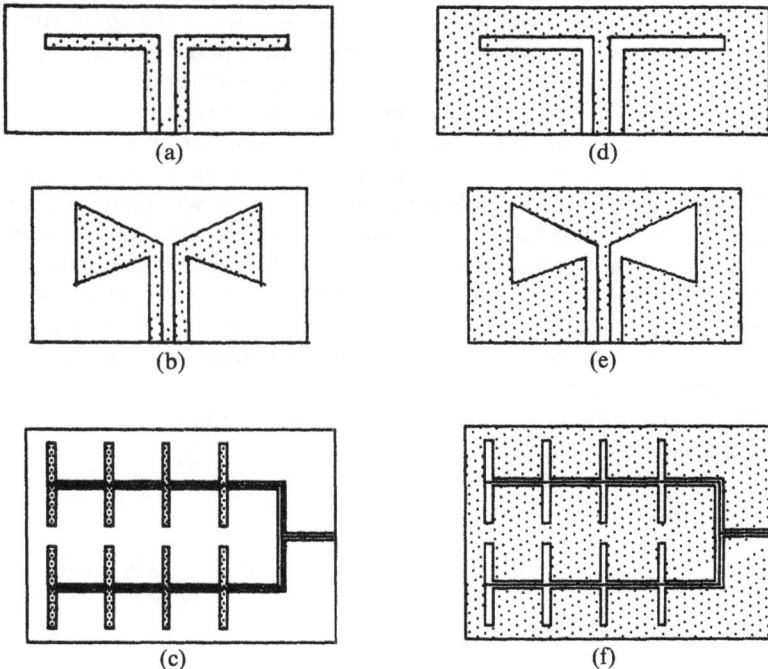

(a)   (d)

(b)   (e)

(c)   (f)

Figure 8.6. Transition of a coplanar line to an antenna: (a) transition of a coplanar strip to a balanced dipole antenna; (b) transition of the coplanar strips into bow-tie strips; (c) an array application of these; and (d–f) corresponding antennas in the slot line.

The advantage of this approach over the microstrip-to-slot-line transition balun presented previously is that this approach is uniplanar and requires no back-plane circuit, thus, it costs less to manufacture. Another advantage of this method is its excellent adaptability for realizing an array feed structure, as shown in Figure 8.6c and 8.6f.

The folded dipole antenna is an extension of the dipole antenna, constructed by folding a dipole antenna at the center of its arms to obtain a folded dipole, as shown in Figure 8.7a. The two conductors will be separated by a distance, b, and their actions will be similar to two symmetric coupled transmission lines that support two distinct modes, the odd and the even. In the even mode, the two conductors will have equal currents and equal voltages, as shown in Figure 8.7c. These even-mode currents will radiate in-phase and result in added antenna gain, thus doubling the far-field radiation. This mode is therefore rightly called the antenna mode.

In contrast to the even mode, the odd mode will have equal currents but they will flow in opposite directions, and equal voltages but with opposite signs, as shown in Figure 8.7c. The radiations of these two currents will cancel out in the far field. For this reason, the mode is a nonradiating mode and, therefore, a transmission-line mode. The input impedance to this antenna will be determined by the impedance of the transmission line and the dipole antenna modes, as we might expect. In practice, the antenna is fed only on one side and the ratio of the terminal voltage to the total terminal current will give the input impedance of this antenna. Referring to Figure 8.7c, the total current and voltage is given by the vector sum of the two currents and voltages on the terminal conductors, as is evident from the corresponding transmission line (odd mode) and antenna mode (even mode) cases.

The input impedance of a shorted transmission line of length $L/2$, with a characteristic impedance of Z_0 ohms, is

$$Z_{\text{In}} = 4Z_xZ_d/(Z_x + 4Z_d) \qquad (8.2)$$

where

$$Z_x = j\,Z_0\,\tan(\beta L/2) \qquad (8.3)$$

and Z_d is the input impedance of an equivalent dipole antenna of length L.

For example, for a folded-wire dipole antenna:

$$Z_0 = 120\,\ln\,(d/a)/\sqrt{\epsilon_r} \qquad (8.4)$$

(a)

(b)

Transmission line mode   Antenna mode

(c)

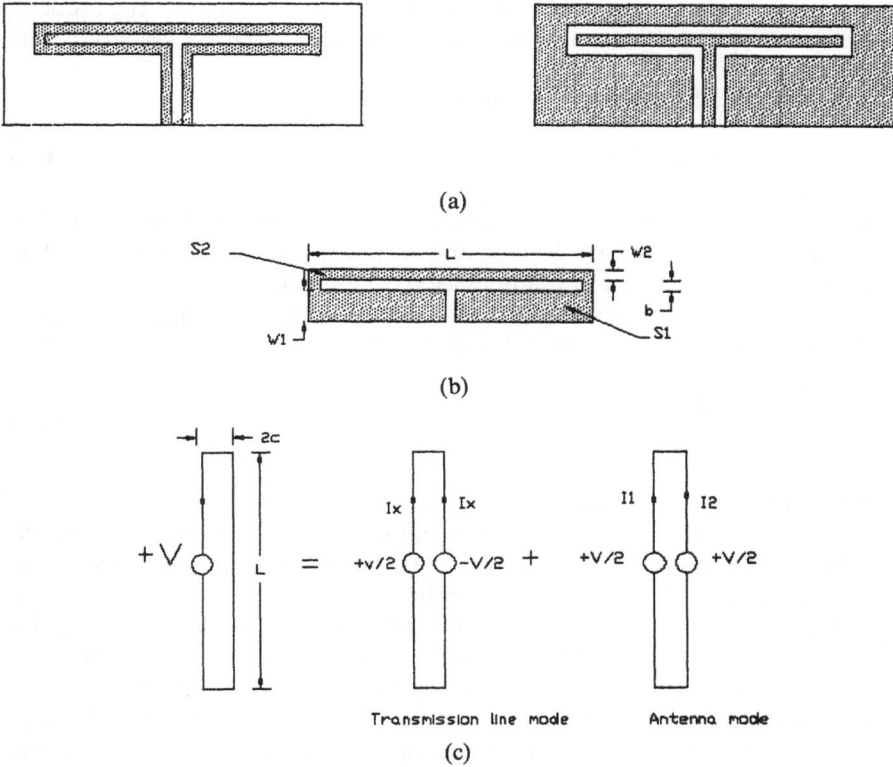

Figure 8.7. (a) **Making folded dipoles from the antennas of Figures 8.6a and 8.6d; (b) an asymmetric folded dipole; (c) the equivalent circuits for transmission line and antenna modes.**

where d is the wire spacing, and a is the wire radius. This impedance for the stripline is more complicated and will be considered later.

The above transmission-line model for the folded dipole has been shown to agree very well with numerical techniques based on the method of moments [2] for the folded-wire dipole antennas. The input impedance for the half-wave folded dipole becomes four times the impedance of the equivalent half-wave dipole dipole as can be seen from (8.2) by letting the input impedance of the shorted quarter-wave transmission line approach infinity.

The above analysis assumed folded dipole antennas of symmetric wire pairs (i.e., wires of the same diameter in the case of wire antennas, and strips or slots of the same width for strip or slot antennas). In practice, different asymmetric conductors are sometimes used to construct the folded dipole.

The input impedance of the folded dipole is

$$Z_{in} = (1 + a)^2 Z_d \tag{8.5}$$

where for thin wires folded dipoles $a = \ln(d/a_1)/\ln(d/a_2)$, and a_1 and a_2 are the wire radii for the derived and folded wires, respectively [2]. The parameters for strip or slot antennas will be given later.

In general, for asymmetric folded dipoles, the input impedance based on the transmission line model is given by

$$Z_{in} = \frac{2(1 + a)^2 Z_x Z_d}{(1 + a)^2 Z_d + 2Z_x} \tag{8.6}$$

where $(1 + a)^2$ is an impedance transformation ratio.

The impedance transformation ratio can be explained by a modification of the antenna mode current. To help visualize this, refer to the coupled line analogy, mentioned earlier, where the odd mode and even mode correspond to the transmission line and the antenna modes. Similarly, the asymmetric folded dipole can be compared to an asymmetric pair of coupled lines, in which the odd mode will stay the same and the even mode will have unequal currents on the two conductors with equal voltages. The parameter a in the equation is the ratio of current of the two conductors of the even mode (the antenna mode).

As mentioned earlier, a printed folded dipole strip or slot can be realized easily on one side of a substrate that can be fed from coplanar strips or coplanar waveguides, as illustrated in Figure 8.7a. The earlier relations for folded dipoles also hold, with the exception that the equivalent dipole impedance and the characteristic impedance of the transmission lines to be used must be those of the corresponding strip or slot dipole and transmission lines, respectively, whatever the case may be.

For an asymmetric coplanar-strip folded dipole the characteristic impedance is found by conformal mapping, given by [6, 7].

$$Z_0 = \frac{120 K(k)}{\sqrt{\epsilon_r} \, K'(k)}$$

$$k = \frac{2b \, [1 + e \, (2b + W_1)]}{2b + W_1 + e(2b)^2}$$

$$e = \frac{W_1 W_2 + 2b \, (W_1 + W_2) - [W_1 W_2 (b + W_1)(b + W_2)]^{1/2}}{(2b)^2 \, (W_1 + W_2)} \tag{8.7}$$

$K'(k) = K(1 - k^2)$. $K(k)$ is the complete elliptic function. Parameter a can be found from a consideration of the total charge on each strip [6]:

$$a = Q_2/Q_1 = \frac{\ln\{4c + 2[(2c)^2 - (W^1/2)^2]^{1/2}\} - \ln(W_1)}{\ln\{4c + 2[(2c)^2 - (W_2/2)^2]^{1/2}\} - \ln(W_2)} \tag{8.8}$$

The equivalent dipole impedance is found by establishing an equivalent circular cylindrical dipole antenna that would exhibit the same per unit length capacitance as the asymmetrical planar antenna under consideration [6]. The relation for the radius of this equivalent cylindrical antenna is

$$p_e = \exp\left(\frac{Q_1}{Q_1 + Q_2}\ln(W_1/4) + \frac{Q_2}{Q_1 + Q_2}\ln(c + [c^2 - (W_2/4)^2]^{1/2})\right) \tag{8.9}$$

where Q_1 and Q_2 are the total charges on strip 1 and 2, respectively. Note that for the symmetric coplanar strips where $W_1 = W_2 = W$, the parameter $k = b/(b + 2W)$. Thus,

$$p_e = \sqrt{(W/4)\{c + [c^2 - (W/4)^2]^{1/2}\}} \tag{8.10}$$

So from these formulas, the equivalent dipole impedance is found from the equivalent cylindrical dipole of radius p_e [6]. The asymmetric coplanar slot antenna are the complementary structure of the asymmetric coplanar strip antenna and its analysis will be very similar. The same conformal transformation can actually be used to obtain the transmission line impedance and the same procedure is used to find the equivalent dipole impedance [6].

8.5 END-FIRE SLOT ANTENNAS

A useful end-fire type of slot antenna can be realized by an open slot, as shown in Figure 8.8. This type of antenna will be a traveling wave as opposed to the standing wave slots introduced earlier; therefore, it will be relatively more broadband and will have a symmetric radiation pattern in the E and H planes. The traveling-wave slot antenna can be thought of as a transition from a slot-line transmission line as the slot line gap is gradually increased to an order of the wavelength over an optimum length of a taper. A typical traveling-wave slot line antenna as it evolved from the slot line is shown in Figure 8.8. The electromagnetic wave is bound to the slot-line transmission line and concentrated in the vicinity of the

132

gap. As the gap is increased, the electromagnetic wave fringes more into the surrounding space. Finally, when the gap is sufficiently large and comparable to the wavelength, the electromagnetic energy is no longer bound to the structure very tightly, and when the structure is suddenly truncated at the edge of the antenna, radiation waves are launched. This mechanism is in the nature of a typical traveling-wave antenna. Figure 8.8 shows a typical slot antenna fed by a microstrip line, realized by a transition from the microstrip to the slot line as considered in Chapter 5. Another possible feed mechanism for the slot antenna is the transition form, a fin line to the slot line that extends out of the waveguide.

The slot line antenna taper can be any suitable realizable taper. An example of an exponentially tapered slot antenna, called Vivaldi, is given by Gibson [8]. A linearly tapered slot antenna similar to that shown in Figure 8.8 is presented in [9]. A thorough study of this class of traveling-wave antennas suitable for microwave integrated circuit is presented and compared by Yngvesson et al. [10].

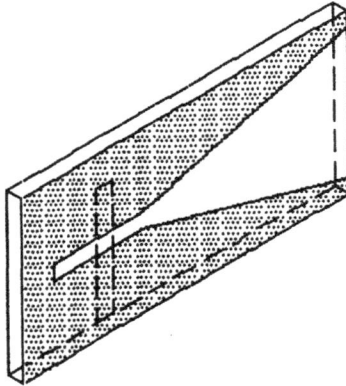

Figure 8.8. An end fire slot antenna as realized from a slot line and fed by a microstrip line.

8.6 MICROSTRIP PATCH ANTENNAS

There are two primary ways of feeding the microstrip patch antennas: direct contact and induction. The first method uses a transmission-line contact either by a microstrip line, as illustrated in the Figure 8.9a [11], or by a coaxial probe from below the substrate, as shown in Figure 8.9b [12]. In the noncontact or the electromagnetic induction method, there is no direct physical or electrical contact between the patch and the feeding structure and the coupling to the antenna element is made strictly by electromagnetic induction, either directly from the feeding microstrip or through an intermediate aperture. This is illustrated in Figure 8.10.

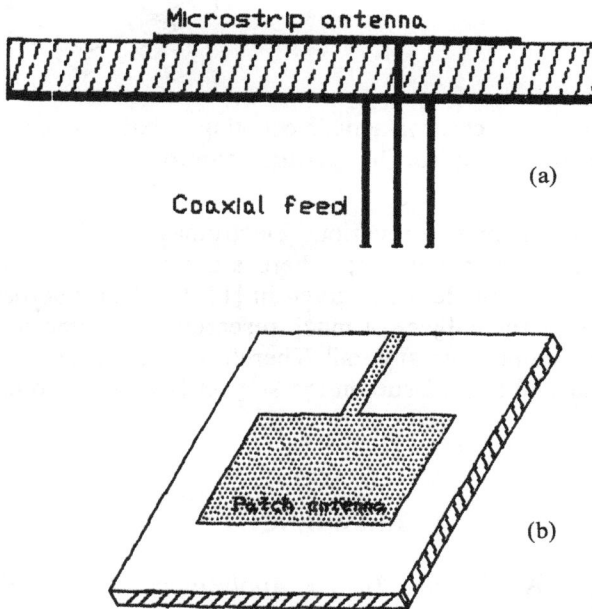

Figure 8.9. Feeding a microstrip antenna by direct connection: (a) coaxial probe, (b) microstrip.

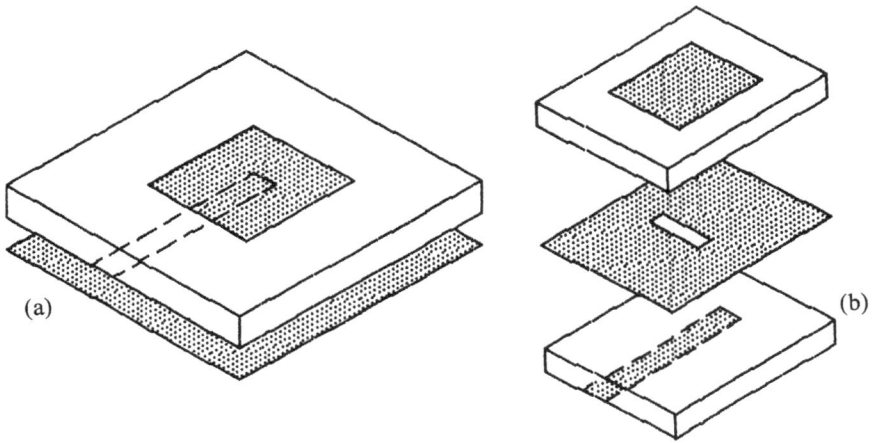

Figure 8.10. A typical electromagnetic coupling of patch antennas: (a) microstrip coupled, (b) aperture coupled.

A tutorial summary of various electromagnetically coupled feed mechanisms can be found in [13], whereas a more elaborate discussion and electromagnetic modeling is given in [14–17]. The microstrip patch antenna has been the subject of much research, and some books cover the subject of feeding in great detail. Therefore, the reader is encouraged to refer to a special focus issue on the subject [18] and a book [19].

REFERENCES

[1] Balanis, C. A., *Antenna Theory, Analysis, and Design*, Harper and Row, New York, 1982.

[2] Stutzman, W. L., and Thiele, G. A., *Antenna Theory and Design*, John Wiley and Sons, New York, 1981.

[3] Richmond, J. H., "Monopole Antenna on Circular Disk," *IEEE Trans. Antennas Propagation*, Vol. AP-32, No. 12, December 1984.

[4] Nesic, A., "Printed Slotted Array Excited by a Coplanar Waveguide," *12th European Microwave Conf. Digest*, 1982.

[5] Axelord, A., and D. Lipman, "Novel Planar Balun Feeds Octave-Bandwidth Dipoles," *Microwaves and RF*, August 1986.

[6] Lampe, R. W., "Design Formulas for an Asymmetric Coplanar Strip Folded Dipole," *IEEE Trans. Antennas Propagation*, Vol. AP-33, No. 9, September 1985.

[7] Gupta, K. C., R. Garg, and I. J. Bahl, *Microstrip Lines and Slotlines,* Artech House, Norwood, MA, 1979.

[8] Gibson, P. J., "The Vivaldi Aerial," *16th European Microwave Conf. Digest,* 1986.

[9] Parsad, S. N., and S. Mahapatra, "A Novel MIC Slot-Line Antenna," *9th European Microwave Conf. Digest,* 1979.

[10] Yngvesson, K. S., D. H. Schaubert, T. L. Korzeniowski, and J. F. Johanson, "Endfire Tapered Slot Antennas on Dielectric Substrates," *IEEE Trans. Antennas Propagation,* Vol. AP-33, No. 12, December 1985.

[11] Davidovitz, M., and Y. T. Lo, "Input Impedance of a Probe-Fed Circular Microstrip Antenna with Thick Substrate," *IEEE Trans. Antennas Propagation,* Vol. AP-34, No. 7, July 1986.

[12] Derneryd, A. G., "Linearly Polarized Microstrip Antennas," *IEEE Trans. Antennas Propagation,* Vol. AP-24, No. 11, November 1976.

[13] Pozar, D. M., "An Update on Microstrip Antenna Theory and Design Including Some Novel Feeding Techniques," *IEEE Antennas Propagation Soc. Newsletter,* October 1986.

[14] Katehi, P. B., and N. G. Alexopoulos, "On the Modeling of Electromagnetic Coupled Microstrip Antennas—The Printed Strip Dipole," *IEEE Trans. Antennas Propagation,* Vol. AP-32, No. 11, November 1984.

[15] Zhang, Q., Y. Fukuoda, and T. Itoh, "Analysis of a Suspended Patch Antenna Excited by an Electromagnetically Coupled Inverted Microstrip Feed," *IEEE Trans. Antennas Propagation,* Vol. AP-33, No. 8, August 1985.

[16] Pozar, D., "A Reciprocity Method of Analysis for Printed Slot and Coupled Microstrip Antennas," *IEEE Trans. Antennas Propagation,* Vol. AP-34, No. 12, December 1986.

[17] Sullivan, P. L., and D. H. Schaubert, "Analysis of an Aperture Coupled Microstrip Antenna," *IEEE Trans. Antennas Propagation,* Vol. AP-32, No. 11, November 1984.

[18] *IEEE Trans. Antennas Propagation,* Vol. AP-29, No. 1, January 1981.

[19] James, J. R., P. S. Hall, and C. Wood, *Microstrip Antenna Theory and Design,* Peter Peregrinus, 1981.

Appendix A
Coaxial Connector Performance

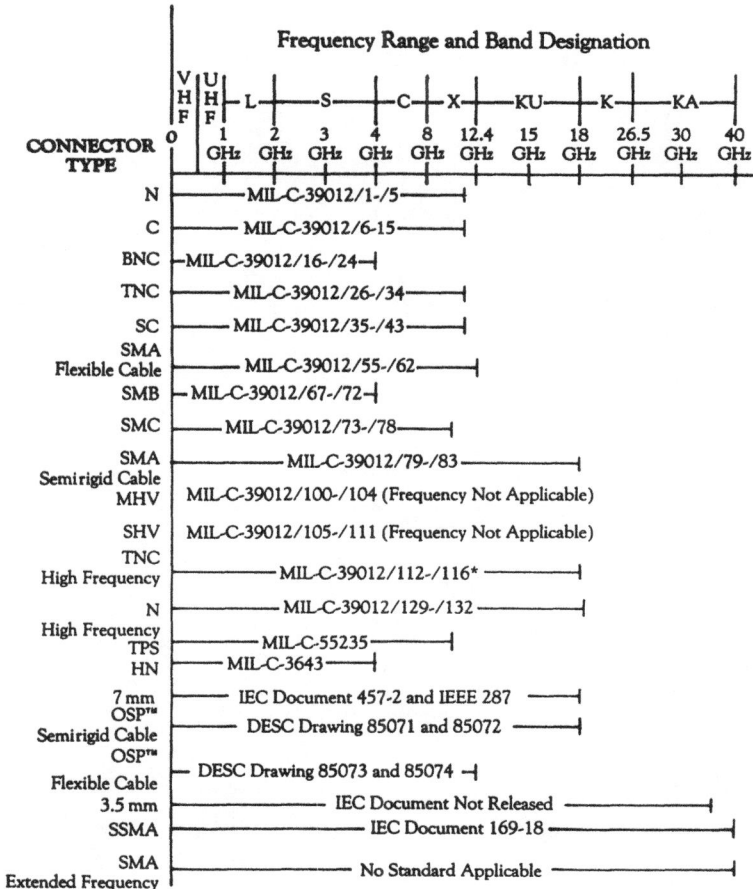

Frequency Range and Band Designation

CONNECTOR TYPE	Standard / Frequency Range
	V H F O | U H F |—L—|—S—|—C—|—X—|—KU—|—K—|—KA—
	1 GHz 2 GHz 3 GHz 4 GHz 8 GHz 12.4 GHz 15 GHz 18 GHz 26.5 GHz 30 GHz 40 GHz
N	MIL-C-39012/1-/5
C	MIL-C-39012/6-15
BNC	MIL-C-39012/16-/24
TNC	MIL-C-39012/26-/34
SC	MIL-C-39012/35-/43
SMA Flexible Cable	MIL-C-39012/55-/62
SMB	MIL-C-39012/67-/72
SMC	MIL-C-39012/73-/78
SMA Semirigid Cable	MIL-C-39012/79-/83
MHV	MIL-C-39012/100-/104 (Frequency Not Applicable)
SHV	MIL-C-39012/105-/111 (Frequency Not Applicable)
TNC High Frequency	MIL-C-39012/112-/116*
N High Frequency	MIL-C-39012/129-/132
TPS	MIL-C-55235
HN	MIL-C-3643
7 mm	IEC Document 457-2 and IEEE 287
OSP™ Semirigid Cable	DESC Drawing 85071 and 85072
OSP™ Flexible Cable	DESC Drawing 85073 and 85074
3.5 mm	IEC Document Not Released
SSMA	IEC Document 169-18
SMA Extended Frequency	No Standard Applicable

*Being revised to 18 GHz

OSP is a trademark of M/A-COM Omni Spectra, Inc.

From *Microwave Systems News Design Handbook.* Copyright 1987 by *Microwave Systems News.* Reprinted with permission.

Appendix B
Connector Glossary

BNC Connector: Features a bayonet coupling, used where quick connect-disconnect is desired yet positive locking is needed.

C Connector: Similar in size to N connectors, however, with bayonet locking. The dielectric overlap provides good voltage handling capabilities, but bayonet coupling does not perform well electrically during vibration.

Coaxial Connector: Comprises two concentric conductors separated by a dielectric (insulator) and is most commonly used as an extension of a transmission line to facilitate interconnections.

Contact: The conducting part of a connector that acts with another such part to complete or break a circuit. Contacts provide a separable through connection in a cable-to-cable, cable-to-module, or module-to-module situation.

Hermaphroditic Connector: An interconnecting device in which both mating parts are identical at their mating surfaces. Also called *Sexless Connectors*.

Hermetic: Airtight. Since all materials are permeable, specifications define acceptable levels of hermeticity. The maximum allowable leak rate for coaxial connectors is usually specified as less than 1×10^{-8} cc/s standard air equivalent with one atmosphere differential.

HN Connector: Slightly larger than N, but uses basically the same cable. It is designed to have increased voltage capabilities without the loss of its RF parameters through about 4 GHz.

Insertion Loss: The loss in load power resulting from the insertion of a connector, component or device.

From *Microwave Systems News Design Handbook*. Copyright 1987 by *Microwave Systems News*. Reprinted with permission.

Instrumentation Grade: A connector designed to give superior RF performance. This normally is a test port connector and is used for precision microwave measurement. A high degree of repeatability can be expected. It is not designed for use in harsh environments.

Interface: The physical connection between two systems or devices or the matching of adjacent components, circuits, or equipment.

MHV Connector: High voltage of the BNC. Ground connection is broken before power connection. Inactivated under MIL-C-39012 and replaced by SHV.

Microstrip: Two parallel conductive planes separated by a dielectric. The narrow strip is the signal conductor and the wider strip is the signal ground.

Military Grade: A connector designed to give adequate RF performance and meet mechanical and environmental requirements usually seen in system applications. It is normally a high usage item.

N Connector: The first matched RF connector exhibiting good RF performance but limited voltage-handling capabilities. It has threaded coupling for mating.

SC Connector: Often referred to as the *Screwed C*. Popular in areas which experience high vibration.

SHV Connector: Designed to replace the MHV for high energy physics applications.

SMA Connector: High performance subminiature connector. Widely used in high performance military systems and test equipment, as well as where miniaturization is desired.

SMB Connector: Snap-on version of the SMC, more limited in its performance.

SMC Connector: Threaded subminiature connector. Does not have the high performance of the SMA.

SSMA Connector: Smaller than SMA for reduced packaging requirements, allows for higher frequency performance.

Stripline: Similar to microstrip except that it has three parallel conductors separated by dielectric. The narrow center strip is the signal conductor and the wider outer strips are the signal ground planes.

TNC Connector: Threaded version of the BNC featuring higher frequency capabilities because of their more stable mating. They are popular where the larger SC and N connectors are too big but a threaded coupling is needed.

TPS Connector: Miniature bayonet connector, smaller than BNC, with improved performance to 10 GHz.

Transmission Line: Used to guide the propagation of electromagnetic waves and confine it to prevent the loss of energy.

Voltage Standing Wave Ratio: (VSWR). The ratio of the incident wave to the reflected wave in a transmission line. This represents a figure of merit for the impedance mismatch within the line and will vary with frequency.

Appendix C
Equations for Embedding and De-embedding Device Measurements

This appendix presents the equations for modifying the error coefficients used inside a network analyzer to provide a measurement vantage point different from that which normal calibration and measurement techniques will allow. Such modification enables the network analyzer to display data as though it had been calibrated at a measurement plane separated from the actual calibration plane by an embedding network.

Also presented are the equations that permit *embedding* the device under test in a hypothetical network to allow viewing the device as though it were actually in a circuit with such a network.

De-embedding Equations

Figures C.1, C.2, and C.3 show the 12 original two-port calibration E terms (error term matrices) labeled $E1[\ ]$–$E4[\ ]$, which are to be cascaded with the fixture error terms, $F1[\ ]$ and $F2[\ ]$, to provide modified terms, E', for use by the network analyzer. It should be noted that the $F1[\ ]$ and $F2[\ ]$ terms correspond to networks physically located on the port 1 and port 2 sides of the DUT, respectively, while the E terms do not have a direct physical representation. This is because the unity $E1[\ ]$ and $E3[\ ]$ forward transmission terms require normalization of the E_{rf}, E_{tf}, E_{rr}, and E_{tr} terms. This normalization causes source side characteristics to be included in the opposite side's transmission tracking term.

Because the de-embedding equations for the forward and reverse directions are symmetrical, only the forward direction is developed here. The reverse direction equations are obtained by appropriate subscript changes of the forward direction equations.

143

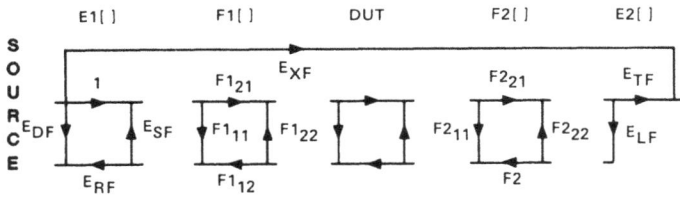

Figure C.1 Two-port forward direction flow diagram.

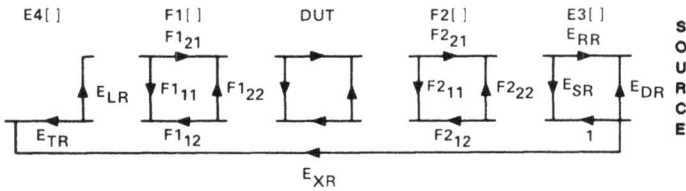

Figure C.2 Two-port reverse direction flow diagram.

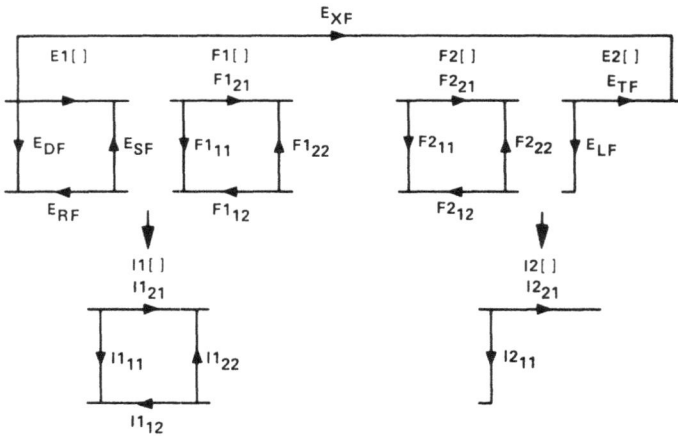

Figure C.3 Forward direction cascading.

The E and F terms on the same side of the DUT are first cascaded using either topographical or analytical methods. The result is an intermediate I matrix on each side of the DUT, $I1[\ ]$ and $I2[\ ]$.

$$I1_{11} = E_{df} + (E_{rf}F1_{11}) / (1 - E_{sf}F1_{11})$$

$$I1_{12} = (E_{rf}F1_{12}) / (1 - E_{sf}F1_{11})$$

$$I1_{21} = (F1_{21}) / (1 - E_{sf}F1_{11})$$

$$I1_{22} = F1_{22} + (E_{sf}F1_{21}F1_{12}) / (1 - E_{sf}F1_{11})$$

$$I2_{11} = F2_{11} + (E_{lf}F2_{21}F2_{12}) / (1 - E_{lf}F2_{22})$$

$$I2_{21} = (E_{tf}F2_{21}) / (1 - E_{lf}F2_{22})$$

It is then necessary to adjust the $I[\ ]$ transmission terms to the form required by the network analyzer internal error model (E terms). This is done by setting $E1'_{21}$ to 1 and placing the product of $I1_{21}$ and $I1_{12}$ into E'_{rf}. Similarly, the forward tracking term, E'_{tf} is obtained by multiplying $I2_{21}$ by this same term, $I1_{21}$. This provides the network analyzer with transmission error terms which are normalized to the source side forward transmission term, $E1'_{21}$.

$$E1'_{21} = 1$$

$$E'_{rf} = E1'_{12} = I1_{12}I1_{21}$$

$$E'_{tf} = E2'_{21} = I1_{21}I2_{21}$$

The other $E'[\ ]$ terms are the same as the $I[\ ]$ terms and the isolation term, E'_{xf}, needs no modification since it is outside of the cascading process.

The resultant 12 two-port de-embedding equations follow:

$$E'_{df} = E_{df} + (E_{rf}F1_{11}) / (1 - E_{sf}F1_{11})$$

$$E'_{rf} = (E_{rf}F1_{12}F1_{21}) / (1 - E_{sf}F1_{11})^2$$

$$E'_{sf} = F1_{22} + (E_{sf}F1_{12}F1_{21}) / (1 - E_{sf}F1_{11})$$

$$E'_{lf} = F2_{11} + (E_{lf}F2_{12}F2_{21}) / (1 - E_{lf}F2_{22})$$

$$E'_{tf} = (E_{tf}F1_{21}F2_{21}) / [(1 - E_{lf}F2_{22}) (1 - E_{sf}F1_{11})]$$

$$E'_{xf} = E_{xf}$$

$$E'_{dr} = E_{dr} + (E_{rr}F2_{22}) / (1 - E_{sr}F2_{22})$$

$$E'_{rr} = (E_{rr}F2_{12}F2_{21}) / (1 - E_{sr}F2_{22})^2$$

$$E'_{sr} = F2_{11} + (E_{sr}F2_{12}F2_{21}) / (1 - E_{sr}F2_{22})$$

$$E'_{lr} = F1_{22} + (E_{lr}F1_{12}F1_{21}) / (1 - E_{lr}F1_{11})$$

$$E'_{tr} = (E_{tr}F1_{12}F2_{12}) / [(1 - E_{lr}F1_{11}) (1 - E_{sr}F2_{22})]$$

$$E'_{xr} = E_{xr}$$

For one-port de-embedding, only the E'_{df}, E'_{rf}, E'_{sf}, or E'_{dr}, E'_{rr}, E'_{sr} are used for port 1 or port 2 measurements, respectively.

Using the De-embedding Algorithm to Embed a Network

The distinction between de-embedding and embedding is made based on the direction of measurement frame movement relative to the DUT. If the network through which the measurement plane moves is real, that is, causal, then movement of the measurement frame toward the DUT is termed de-embedding while movement away from the DUT is called embedding. If no restriction is placed on the network to be considered then the distinction between de-embedding and embedding based on direction of measurement frame movement disappears. It then is possible to use either algorithm for both the de-embedding and the embedding processes.

To use the de-embedding equations to embed a network, some "preconditioning" of the network to be embedded must first take place. Consider the following cascaded networks:

In Figure C.4, $N[\ ]$ is intended to represent the network which is to be embedded while $N\#[\ ]$ is the corresponding "antinetwork." Since the cascade of these two networks is defined to be an identity network, it can be seen that one of the two will always be causal while the other is noncausal, except for the particular case where both are identity networks.

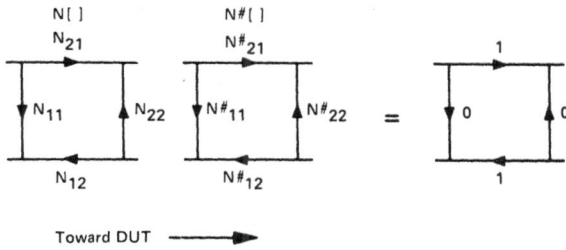

Figure C.4 Cascade of network and antinetwork.

Now consider a two-tier de-embedding process of these cascaded networks. It is clear that the result must be an unchanged measurement frame. After the first tier of de-embedding, the measurement frame has moved through the $N[\ ]$ network and effectively de-embedded the measurement from it. After the second tier of de-embedding, the measurement frame will again have moved, this time through the antinetwork, $N\#[\ ]$. The measurement frame is now on the right side of the cascaded networks, but because the cascade is defined to be an identity network, this is the same as a measurement frame on the left side of the cascade. Movement of the measurement frame toward the DUT (de-embedding) through the antinetwork, $N\#[\ ]$, is therefore the same as movement of the measurement frame away from the DUT (embedding) through the network, $N[\ ]$.

In order to embed a network, $N[\ ]$, it is only necessary to calculate and de-embed the antinetwork $N\#[\ ]$.

To calculate the antinetwork, first write the cascading equations for the two networks:

$$S_{11} = N_{11} + (N_{21}N_{12}N\#_{11}) / (1 - N_{22}N\#_{11})$$

$$S_{12} = (N_{12}N\#_{12}) / (1 - N_{22}N\#_{11})$$

$$S_{21} = (N_{21}N\#_{21}) / (1 - N_{22}N\#_{11})$$

$$S_{22} = N\#_{22} + (N\#_{12}N\#_{21}N_{22}) / (1 - N_{22}N\#_{11})$$

Setting the resulting matrix, $S[\ ]$, equal to an identity matrix,

$$S_{11} = S_{22} = 0 \text{ and } S_{21} = S_{12} = 1$$

gives

$$0 = N_{11} + (N_{21}N_{12}N\#_{11}) / (1 - N_{22}N\#_{11})$$

$$1 = (N_{12}N\#_{12}) / (1 - N_{22}N\#_{11})$$

$$1 = (N_{21}N\#_{21}) / (1 - N_{22}N\#_{11})$$

$$0 = N\#_{22} + (N\#_{12}N\#_{21}N_{22}) / (1 - N_{22}N\#_{11})$$

Solving for $N\#_{11}$ gives

$$N\#_{11} = N_{11} / (N_{11}N_{22} - N_{21}N_{12})$$

which may be used to solve

$$N\#_{12} = (1 - N_{22}N\#_{11}) / N_{12}$$

$$N\#_{21} = (1 - N_{22}N\#_{11}) / N_{21}$$

which may be used to get

$$N\#_{22} = (N\#_{12}N\#_{21}N_{22}) / (N_{22}N\#_{11} - 1)$$

To embed a network, $N[\ ]$, first calculate the antinetwork, $N\#[\ ]$, using the above equations and then provide the results to the de-embedding routine as the network to be de-embedded. De-embedding of this antinetwork, $N\#[\ ]$ is the same as embedding the original network $N[\ ]$.

Index

Analyzer
  automatic network (ANA), 96
  calibration, 102
  vector automatic network (VANA), 81
Antenna, 119
  end-fire slot, 131
  microstrip patch, 133
  monopole, 124
  printed, 126
  transitions, 120
Antipodal finlines
  approach, 62
  test fixtures, 90

Bias network, dc, 96

Calibration
  automatic network analyzer, 102
  test fixture, 96
Chip carrier design, 93
Circuit, coaxial-to-microstrip, 22
Coaxial test fixtures, 83
Coplanar waveguide, 66
  see also Transitions

De-embedding, 101
  and characterization of S-parameters, 19
  and fixture characterization, 107
  in the time domain, 110
Device modeling, 112
Device under test (DUT), 19

Element values deduced
  insertion loss, 26
  measured return loss, 22
Equivalent circuit, 22

Field matching, 2

Gating, 112
Geometric theory of diffraction (GTD), 123

Impedance, 4
  modal wave, 5
  taper, 8
  waveguide, 5

Launchers, waveguide
  electric probe, 38
  magnetic loop, 49
Loss
  insertion, 26
  measured return, 22

Measurement
  open, short, and load (OSL), 105
  through, reflection, and line (TRL), 106
  through, short, and delay (TSD), 105
Microwave
  integrated circuits (MIC), 15
  monolithic MIC (MMIC), 113
  test fixtures, 81

Perfect electric conductor (PEC), 2
Perfect magnetic conductor (PMC), 2

Ridged waveguide
  approach, 53
  test fixtures, 93

Slotline transitions, 65
S-parameters, 19

Test fixtures
  antipodal fin-line, 90
  coaxial, 83
  evaluation and calibration, 96
  microwave, 81

149